Solutions and Tests For Exploring Creation With Physics

© 1998, 1999, 2000 Apologia Educational Ministries, Inc.
All rights reserved.

Manufactured in the United States of America
Sixth Printing 2003

Published By
Apologia Educational Ministries, Inc.
Anderson, IN

Printed by
The C.J. Krehbiel Company
Cincinnati, OH

Exploring Creation With Physics
Solutions and Tests

TABLE OF CONTENTS

Teacher's Notes 1

Answers to the Review Questions

Answers to the Review Questions for Module #1 6
Answers to the Review Questions for Module #2 7
Answers to the Review Questions for Module #3 8
Answers to the Review Questions for Module #4 9
Answers to the Review Questions for Module #5 10
Answers to the Review Questions for Module #6 11
Answers to the Review Questions for Module #7 12
Answers to the Review Questions for Module #8 13
Answers to the Review Questions for Module #9 14
Answers to the Review Questions for Module #10 15
Answers to the Review Questions for Module #11 16
Answers to the Review Questions for Module #12 17
Answers to the Review Questions for Module #13 18
Answers to the Review Questions for Module #14 19
Answers to the Review Questions for Module #15 21
Answers to the Review Questions for Module #16 22

Answers to the Practice Problems

Answers to the Practice Problems for Module #1 26
Answers to the Practice Problems for Module #2 29
Answers to the Practice Problems for Module #3 32
Answers to the Practice Problems for Module #4 37
Answers to the Practice Problems for Module #5 44
Answers to the Practice Problems for Module #6 54
Answers to the Practice Problems for Module #7 61
Answers to the Practice Problems for Module #8 68
Answers to the Practice Problems for Module #9 74
Answers to the Practice Problems for Module #10 81
Answers to the Practice Problems for Module #11 86
Answers to the Practice Problems for Module #12 92
Answers to the Practice Problems for Module #13 97
Answers to the Practice Problems for Module #14 103
Answers to the Practice Problems for Module #15 113

Tests

Test for Module #1 121
Test for Module #2 123
Test for Module #3 125
Test for Module #4 127
Test for Module #5 129
Test for Module #6 131
Test for Module #7 133
Test for Module #8 137
Test for Module #9 139
Test for Module #10 141
Test for Module #11 143
Test for Module #12 145
Test for Module #13 147
Test for Module #14 151
Test for Module #15 153
Test for Module #16 155

Answers to the Tests

Answers to the Module #1 Test 158
Answers to the Module #2 Test 161
Answers to the Module #3 Test 164
Answers to the Module #4 Test 168
Answers to the Module #5 Test 173
Answers to the Module #6 Test 180
Answers to the Module #7 Test 186
Answers to the Module #8 Test 192
Answers to the Module #9 Test 197
Answers to the Module #10 Test 203
Answers to the Module #11 Test 208
Answers to the Module #12 Test 213
Answers to the Module #13 Test 216
Answers to the Module #14 Test 221
Answers to the Module #15 Test 228
Answers to the Module #16 Test 232

TEACHER'S NOTES

Exploring Creation With Physics

Thank you for choosing *Exploring Creation With Physics*. I designed this modular course specifically to meet the needs of the homeschooling parent. I am very sensitive to the fact that most homeschooling parents do not know physics very well, if at all. As a result, they consider it nearly impossible to teach to their children. This course has several features that make it ideal for such a parent.

1. The course is written in a conversational style. Unlike many authors, I do not get wrapped up in the desire to write formally. As a result, the text is easy to read and the student feels more like he or she is *learning*, not just reading.

2. The course is completely self-contained. Each module includes the text of the lesson, experiments to perform, problems to work, questions to answer, and a test to take. The solutions to the problems and questions are fully explained, and the test answers are provided. The experiments are written in a careful, step-by-step manner that tells the student not only what he or she should be doing, but also what he or she should be observing.

3. The materials for the experiments are all readily available at either the grocery or hardware store. In addition, all of the experiments can be performed with household equipment such as balls, stopwatches, boards, and springs.

4. Most importantly, this course is Christ-centered. In every way possible, I try to make the science of physics glorify God. One of the most important things that you and your student should get out of this course is a deeper appreciation for the wonder of God's creation!

I hope that you and your student enjoy taking this course as much as I have enjoyed writing it.

Pedagogy of the Text

(1) There are three types of exercises that the student is expected to complete: "on your own" problems, review questions, and practice problems.

- The "on your own" problems should be solved as the student reads the text. The act of working out these problems will cement in the student's mind the concepts he or she is trying to learn. The solutions to these problems are included as a part of the student's text. The student should feel free to use these solutions to help understand the problems.

- The review questions are conceptual in nature and should be answered after the student completes the module. They will help the student recall the important concepts from the reading. As your student's teacher, you can decide whether or not they can look at the solutions to these questions. They are located in this book.

- The practice problems should also be solved after the module has been completed, allowing the student to review the important quantitative skills from the module. As your student's teacher, you can decide whether or not they can look at the solutions to these problems. They are located in this book.

(2) In addition to the problems, there is also a test for each module in this book. **I strongly recommend that you administer each test once the student has completed the module and all associated exercises. The student should be allowed to have only a calculator, pencil, and paper while taking the test.** I understand that many homeschoolers do not like the idea of administering tests. However, if your student is planning to attend college, it is *absolutely* necessary that he or she become comfortable with taking tests!

(3) Any information that the student must memorize is centered in the text and put in boldface type. In addition, all definitions presented in the text need to be memorized. Finally, if an equation must be used to answer any "on your own" problem, practice problem, or review question, then it must be memorized for the test. In general these student exercises are meant as a study guide for the tests. Skills and knowledge necessary to complete these student exercises will be required for the test.

(4) Words that appear in bold-face type (centered or not) in the text are important terms that the student should know.

(5) The equations are numbered so that I can refer to them easily.

(6) When looking at the solutions to the students exercises and tests, you will notice that every solution contains an underlined section. That is the answer. The rest is simply an explanation of how to get the answer. For questions that require a sentence or paragraph as an answer, the student need not have *exactly* what is in the solution. The basic message of his or her answer, however, has to be the same as the basic message given in the solutions.

Experiments

The experiments in this course are designed to be done as the student is reading the text. I recommend that your student keep a notebook of these experiments. This notebook serves two purposes. First, as the student writes about the experiment in the notebook, he or she will be forced to think through all of the concepts that were explored in the experiment. This will help the student cement them into his or her mind. Second, certain colleges might actually ask for some evidence that your student did, indeed, have a laboratory component to his or her physics course. The notebook will not only provide such evidence but will also show the college administrator the quality of the physics instruction that you provided to your student. I recommend that you perform your experiments in the following way:

- When your student gets to the experiment during the reading, have him or her read through the experiment in its entirety. This will allow the student to gain a quick understanding of what her or she is to do.

- Once the student has read the experiment, he or she should then start a new page in his or her laboratory notebook. The first page should be used to write down all of the data taken during the experiments and perform any calculation explained in the experiment.

- When the student has finished the experiment, he or she should write a brief report in his or her notebook, right after the page where the data and calculations were written. The report should be a brief discussion of what was done and what was learned. The report should be written so that someone who had never read the experiment in the text could understand the basics of what was done and what was learned. It needn't be incredibly detailed, but it should be written clearly and with good grammar.

- **PLEASE OBSERVE COMMON SENSE SAFETY PRECAUTIONS. The experiments are no more dangerous than most normal, household activities. Remember, however, that the vast majority of accidents do happen in the home. Chemicals should never be ingested; hot beakers and flames should be regarded with care; and OSHA recommends that all physics experiments be performed while wearing some sort of eye protection such as safety glasses or goggles.**

Question/Answer Service

For all those who use my curriculum, I offer a question/answer service. If there is anything in the modules that you do not understand - from an esoteric concept to a solution for one of the problems - just contact me by any of the means listed on the **NEED HELP?** page located at the front of the student text.

Dr. Jay L. Wile

A Word About Grading

The physical sciences are, by far, the most difficult of all subjects to study. As a result, students often perform significantly worse in courses like chemistry and physics than they do in all other subjects, including math. This often makes the student feel that he or she is not talented in the sciences, because that's where the student gets his or her lowest grades. Often, however, this is not the case. Some of the best chemists and physicists I know received lower grades in chemistry and physics than in any of their other courses. In fact, my own *lowest* GPA in college was my chemistry GPA. Thus, just because a straight-A student gets B's or C's in this course, they should not be discouraged from taking more physical science courses.

Public schools have long recognized this fact, so they implement strategies that tend to "boost" their students' grades in the physical sciences. For example, all public schools give their students grades on their labs and homework. Since labs and homework are always performed with the help of the teacher and fellow students, the grades on these assignments are usually quite high. This tends to boost the lower test scores, allowing students to have grades that are comparable to their other courses. Since all public schools do this, and since college admissions people (or job interviewers) will be comparing your student to publicly-schooled students, you should probably do the same. Here are my suggestions on how to grade your students:

1. Give them a grade for each lab that they do. This grade should not reflect the accuracy of the student's results. Rather, it should reflect how well the student followed directions and how well he or she wrote up the lab in his or her lab notebook.

2. Give them a grade for each test. If a test problem contains multiple parts, it should be worth more points than other test questions that do not. As a general rule, I would say that every answer that a student must write down is worth one point. That way, their percentage grade can be calculated as total number of correct answers divided by the total number of answers given. Additionally, you can give partial credit. If a student plowed through the entire problem correctly but just messed up on the calculator, the student should receive 3/4 of a point. If the student got the first couple of steps correct and messed up after that, they should receive 1/2 of a point. Of course, this grading technique requires that you learn the subject right along with the student. This is, of course, what I recommend that you do to begin with!

3. The student's overall grade in the course should be weighted as follows: 35% lab grade, 65% test grade. A straight 90/80/70/60 scale should be used to calculate the student's letter grade. This is typical for most public schools.

Finally, I must tell you that I pride myself on the fact that this course is user-friendly and reasonably understandable. At the same time, however, *it is not EASY*. This is a tough course. I have designed it so that any student who gets a "C" or better on the tests will be VERY well prepared for college.

Solutions To The

Review Questions

MODULE 1 - ANSWERS TO THE REVIEW QUESTIONS

1. Pretty much everything except light contains matter. Since (d) is the only thing on the list that is just light, then (d) lightning bolt has no matter.

2. Length is measured in meters, mass in grams, volume in liters, and time in seconds.

3. The prefix "centi" means 0.01, or one hundredth.

4. Since a kL is equal to 1,000 L and a mL is only equal to 0.001 L, the glass holding 0.5 kL has more liquid in it.

5. 2.65 cm (NOTE: Numbers from 2.63 cm to 2.67 cm are acceptable.)

6. Assuming both students reported the proper number of significant figures, the first student was more precise, but the second student was more accurate. Remember, precision is determined by the significant figures used. Accuracy, however, is determined by how close you are to the *correct* answer.

7. a. 6
   b. 5
   c. 2
   d. 3

8. The student's value for density has far too many significant figures and it has no units attached. Either of these things would make the student's answer wrong.

9. Ice floats on water because its density is lower than that of water. As we learned in Experiment 1.4, a substance floats on top of another substance if its density is lower.

10. The golden statue would be heavier because its density is greater.

MODULE 2 - ANSWERS TO THE REVIEW QUESTIONS

1. A vector quantity has information about direction, while a scalar quantity does not.

2. Both students got the right answer, but they defined their directions in the opposite way. Since their directions were defined differently, their answers had the opposite signs. The way to avoid this confusion is to explicitly state the direction in reference to a fixed point. For example, 2.3 m/sec^2 towards the building is better than simply -2.3 m/sec^2.

3. Velocity is the vector quantity.

4. Average velocity uses a large time interval in Equation (2.1). Instantaneous velocity uses an infinitesimally small time interval.

5. The slope of a displacement versus time graph is the velocity.

6. Physicists say that velocity is relative because the velocity an observer sees actually depends on the velocity of that observer.

7. Since the units attached to the number are m/sec^2, the only possible physical quantity it could be is acceleration, since that is the only physical quantity with those units.

8. Since velocity and acceleration have opposite signs, the acceleration is opposing the velocity. Thus, the object is slowing down.

9. Since acceleration tells us how velocity changes with time, you need to study velocity versus time graphs to learn about acceleration.

10. No. Just because velocity is zero, acceleration does not have to be zero. Acceleration is the *change in* velocity. Thus, if an object is turning around, its velocity changes sign. Through this process, the velocity is briefly zero. Since there is still acceleration, however, the velocity can continue to change and thus the object can complete the turn.

MODULE 3 - ANSWERS TO THE REVIEW QUESTIONS

1. In order to be able to use the three equations we derived in the module, motion must always occur in a straight line and the acceleration of the object must be constant.

2. Air resistance slows down the acceleration of objects that are falling. It arises because an object that falls must shove the molecules and atoms which make up the air out of the way in order to fall. Those molecules and atoms resist the shoving, which slows the object's acceleration.

3. Free fall is the unobstructed fall of an object under the influence of gravity.

4. In fact, air resistance is an obstruction. Since all objects falling near the surface of the earth are in air, they all experience obstruction. Since free fall, by definition, cannot have obstruction, no object can technically experience free fall.

5. Even though objects experiencing air resistance cannot technically experience free fall, many objects are not strongly affected by air resistance. As a result, the air resistance is negligible and can be ignored.

6. The object's velocity is zero at the maximum height that it attains. Its acceleration is *always* 9.8 m/sec^2 downwards, even when its velocity is zero.

7. Since gravity is weaker on the moon, it cannot accelerate things as quickly. Thus, the acceleration due to gravity on the moon is smaller than that of earth.

8. Since air has been removed, there is no air resistance. They are both only affected by gravity, which affects all objects the same. Thus, they both fall at exactly the same rate.

9. As discussed in the text, when the ball returns to the place it was thrown, it has equal and opposite velocity. Thus, its velocity is -1.2 m/sec, or 1.2 m/sec down.

10. Since the parachute increases air resistance, the velocity at which air resistance cancels out all acceleration occurs at a very low velocity. In other words, by increasing air resistance, the parachute decreases the parachutist's terminal velocity to a reasonably safe value.

MODULE 4 - ANSWERS TO THE REVIEW QUESTIONS

1. The angle of a vector is always defined relative to the positive x-axis.

2. a. Speed is a measure of how fast. That's the magnitude of a vector.

   b. Heading is another word for direction. That's the angle of a vector.

   c. Direction is the angle of a vector.

   d. Distance is a measure of how far. That's the magnitude of a vector.

3. All vectors have an x- and y-component. The x-component of this particular vector is zero, but it is there, as it is in all vectors.

4. When the x-component is negative, it means that the vector lies to the left of the origin. A positive y-component means it lies above the origin. The region to the left and above the origin is region II. In this region, we add 180.0 degrees to the result of Equation (4.3) in order to make sure that the angle is defined properly.

5. When the x-component is negative, it means that the vector lies to the left of the origin. A negative y-component means it lies below the origin. The region to the left and below the origin is region III. In this region, we add 180.0 degrees to the result of Equation (4.3) in order to make sure that the angle is defined properly.

6. When the x-component is positive, it means that the vector lies to the right of the origin. A negative y-component means it lies below the origin. The region to the right and below the origin is region IV. In this region, we add 360.0 degrees to the result of Equation (4.3) in order to make sure that the angle is defined properly.

7. If the vector were straight up the positive y-axis, its angle would be 90°. If it were straight across the negative x-axis, it would be 180.0°. This vector seems the be halfway between the two, so its angle is about 135°.

8. Vector **B** has a larger magnitude, indicating that object #2 is accelerating more quickly.

9. In order to be speeding up, the velocity and acceleration vectors must be pointed in the same directions. This is the case for object #1.

10. two one-dimensional

MODULE 5 - ANSWERS TO THE REVIEW QUESTIONS

1. You should always try to see the one-dimensional problems that make up the two-dimensional situation.

2. When a projectile is at its maximum height, the y-component of its velocity is zero. The x-component is the same as when the projectile was launched.

3. For a projectile of this type, the maximum height is always reached at the midpoint of the journey.

4. Since the x-component of the velocity never changes and the magnitude of the y-component when it lands is the same as the magnitude when it is launched, the overall speed is the same when it lands as it was when it was launched. Thus, the speed is 150 m/sec.

5. When the projectile is first launched, the y-component of its velocity continually decreases until it reaches zero. After that, the y-component becomes more and more negative until the projectile hits the ground.

6. No. Since there is no gravity in space, there will be no acceleration in the y-dimension. There is always no acceleration in the x-dimension. As a result, there is no acceleration in either dimension. This means the velocity will never change, so the projectile will travel in a straight line.

7. B. Only in this situation does the projectile land at a height equal to that from which it was launched.

8. This is the same situation as the fired bullet/dropped bullet I asked you about earlier. If the man drops the can, it will begin to accelerate downwards with the acceleration due to gravity. The same thing will happen to the bullet once the gun is fired. As a result, the bullet and the can (in the y-dimension) are following identical paths. Thus, they will both fall the same amount in y-dimension. Therefore, the sharpshooter needs to aim directly at the can.

9. In this situation, the can will not fall in the y-dimension, because the man will hold on to it. The bullet, however, will fall as it travels to the can. Thus, to hit the can, the sharpshooter must aim above the can, to account for the fact that the bullet will fall a little.

10. We have neglected air resistance. That will change the acceleration in the y-dimension as well as add an acceleration to the x-dimension. However, as long as the projectiles we deal with are dense, it is okay to ignore this.

MODULE 6 - ANSWERS TO THE REVIEW QUESTIONS

1. First Law - An object at rest or in motion stays in that same state until acted on by an outside force.

Second Law - The sum of the forces applied to an object is equal to the object's mass times its acceleration.

Third Law - For every applied force there is an equal and opposite force.

2. All of matter is composed of atoms and, on the atomic scale, no surface is smooth. Thus, there will always be some bumps and grooves in any surface, and those bumps and grooves cause friction.

3. The man will fly forward, over the horse's head. The Law of Inertia tells us that an object in motion (the man) stays in motion until acted on by an outside force. The man is moving with the velocity of the horse. When the horse stops, the man is still moving. The force of friction between the man and his saddle is not enough to stop the motion, so the man cannot stop moving forward.

4. The less massive object will accelerate 1,000 times faster than the more massive object.

5. The pound is one of our force units. The $\frac{slug \cdot mile}{hr^2}$ also works because it is a mass unit times a distance unit divided by a time unit squared.

6. The student was measuring weight, because that is the quantity with the unit Newtons.

7. The mass will not be accurate at one of the locations. The scale must measure weight, and it converts to mass by dividing by the acceleration due to gravity. The acceleration due to gravity depends on position, however. Thus, if the number that the scale divides by is right for one of the locations, it will not be right for the other.

8. The normal force is the force exerted by a surface supporting an object. The force counteracts gravity so that the supported object does not fall. It also helps determine the frictional force.

9. The static coefficient of friction is always greater. Thus, 0.54 is the static coefficient and the other one is the kinetic coefficient.

10. Newton's Third Law says that the road must respond by pushing the tire. This is, in fact, why a car moves down the road. The tires push backwards on the road. In response, the road pushes forward (equal and opposite), which makes the car go forward.

MODULE 7 - ANSWERS TO THE REVIEW QUESTIONS

1. For an object to be in translational equilibrium, the sum of the forces acting upon it must be zero.

2. In dynamic equilibrium, an object is moving, but it has no acceleration. In static equilibrium, there is also no acceleration, but the object does not move at all.

3. Tension is the force from a tight string, rope, or chain. Since it is a force, it is measured in Newtons.

4. Situation B results in more tension. In situation B, the tension in each string has an x-component as well as a y-component. The x-components of the two tensions oppose each other. Thus, since the strings fight each other in the x-dimension as well as gravity in the y-dimension, this results in more tension. In situation A, the strings have no x-components to their tensions. Thus, they do not fight each other. They only fight gravity. Since they are fighting fewer forces, there is less tension.

5. Force is the impetus required for translational motion. Torque (which is force times lever arm) is the impetus required for rotational motion.

6. For an object to be in rotational equilibrium, the sum of the torques acting on it must be zero.

7. Situation A results in more torque, because the force is applied perpendicular to the lever arm. Only the component of the force perpendicular to the lever arm counts towards the torque. Since the force is perpendicular to the lever arm in situation A, then all of the force counts towards the torque. In situation B, the force is slanted relative to the lever arm, so only a part of it (the perpendicular component) can count towards the torque.

8. The mechanic should get a longer wrench. Increasing the lever arm will increase the torque.

9. The bigger the car, the harder it is to turn the wheels. If the steering wheel is large, then the driver grasps the wheel far from the axis of rotation. This increases the lever arm, allowing the driver to apply more torque to the steering wheel, making it a little easier to turn.

10. The motion on an incline always occurs parallel to the incline, and the normal force exerted by the incline is perpendicular to the incline. Thus, these dimensions are the best to work in, because that's where all of the interesting action is.

MODULE 8 - ANSWERS TO THE REVIEW QUESTIONS

1. The acceleration is not necessarily zero. In circular motion, the speed is constant, but because the direction is always changing, the velocity changes. Since velocity changes, there is a non-zero acceleration.

2. Since the object is moving on the arc of a circle, it is in circular motion. Thus, the acceleration vector must be pointed directly into the center of the circle, while the velocity vector must be tangent to the circle:

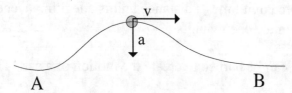

3. Situation A results in uniform circular motion because force and velocity are perpendicular to one another.

4. According to Equation (8.2), the centripetal acceleration depends on the speed squared. If the speed doubles, the centripetal acceleration quadruples.

5. The student is incorrect. The objects do not move towards one another because, in this case, gravity is not strong enough to overcome friction.

6. The reaction force is the gravitational force that the second object exerts on the first. Since the gravitational attraction is mutual, each object exerts the same magnitude of force on each other. The directions are opposite, in accordance to Newton's Third law.

7. Equation (8.3) tells us that the gravitational force decreases the farther you are away from the center of the earth. Thus, the gravitational acceleration will be greater at sea level, because that altitude is closer to the center of the earth.

8. Equation (8.3) says that the gravitational force depends inversely on the square of the distance between the objects' centers. Thus, if the distance doubles, the gravitational force decreases by a factor of four.

9. If the fly increases its speed, it travels around the circle faster. Thus, its orbital period decreases. Since frequency and period are inversely related, this means that orbital frequency increases.

10. If the sun's mass were larger, then the gravitational force it would exert on all of the planets would be greater. This means that the planets would be attracted towards the sun with a larger force. Thus, the planets would orbit closer to the sun.

MODULE 9 - ANSWERS TO THE REVIEW QUESTIONS

1. <u>The secret agent has done no work, because no motion occurred.</u> Work cannot occur without motion.

2. <u>The man on the right will do more work, because the force he is applying is parallel to the motion.</u> The man on the left wastes the component of force which is perpendicular to the motion.

3. Any unit that has a force unit times a distance unit is a legitimate energy unit. Thus, <u>inch·lbs and Newton·km</u> are legitimate energy units.

4. a. A tank of gasoline can be burned to create the motion of a car. Thus, it has <u>potential energy</u>.
   b. A car traveling down the road is in motion, meaning its energy is <u>kinetic</u>.
   c. The electricity is moving, meaning its energy is <u>kinetic</u>
   d. The apple could fall, thus, the apple has <u>potential energy</u>. It is also a source of food, so the chemicals in the apple also have potential energy.

5. <u>The rock's potential energy is greatest when its height is the greatest; thus, at the very beginning of the fall. Its kinetic energy is the greatest when all of the potential energy has been converted to kinetic. This occurs at the instant before the rock hits the ground.</u> Since the rock's total energy is equal to its initial potential energy, it will have an equal amount of potential and kinetic energy when half of its initial potential energy has been converted to kinetic. <u>That occurs halfway down.</u>

6. In lifting the box, you are working on it. As you raise it, you are increasing its potential energy. Once it is on the table, its kinetic energy has not changed, but its potential energy has increased. <u>Thus, you have added potential energy to the box.</u>

7. As was shown in Example 9.7, friction works against the motion of a car, taking energy away from it. The engine must burn gasoline in order to replace this lost energy. The more friction, the more energy must be replaced. <u>Lighter cars get better gas mileage because the frictional force between them and the road is less.</u>

8. <u>Energy is the ability to do work, while power is a measure of how much work was done in a specific time interval.</u>

9. <u>The second light bulb is brighter, meaning it has a larger Wattage. This means that the second bulb consumes more power.</u>

10. Any energy unit divided by a time unit is a legitimate power unit. Thus, $\frac{\text{kiloJoules}}{\text{year}}$ <u>is a legitimate power unit.</u> Also, since Watt is a power unit, any prefix to Watt is also valid. Thus, <u>kiloWatts is also valid.</u>

MODULE 10 - ANSWERS TO THE REVIEW QUESTIONS

1. Momentum and velocity have exactly the same direction. Thus, the momentum vector also has a direction of 45 degrees northeast.

2. Any mass unit times distance unit divided by time unit could be a momentum unit, because that's the unit you get when you multiply mass times velocity. Thus, $\frac{g \cdot km}{min}$ and $\frac{ft \cdot slug}{hr}$ are both legitimate.

3. Since momentum is mass times velocity, the tiny car can have the same momentum as the heavy car as long as it is traveling with a larger velocity.

4. This is a trick question. The wall exerts the same impulse on each. Remember, impulse is equal to the change in momentum. Since both the ball and the clay have the same mass and velocity, they have the same momentum. They also both come to a halt. Thus, they have the same change in momentum. This means the wall must exert the same impulse on both.

5. Even though the wall imparts the same impulse to both objects, the wall exerts more force on the golf ball. This is because the golf ball cannot flex like the clay. Thus, the wall has less time in which to stop the ball. To impart the same amount of impulse, then, the wall must use more force.

6. Follow-through allows the ball and the racquet to be in contact for a longer time interval. This increases the impulse, which increases the change in momentum of the ball.

7. Momentum is only conserved when the sum of the forces acting on a system is zero.

8. Since everything else is equal, if the first fires heavier bullets, those bullets will have more momentum. Since they have more momentum, the first rifle will have to recoil with greater momentum. This means that the first gun will have the greater recoil velocity.

9. A rocket is launched because, as its fuel burns, the combustion products of that fuel speed out under the rocket. These combustion products have momentum. In order to conserve momentum, then, the rocket must travel in the opposite direction.

10. Angular momentum is conserved when the sum of the torques on a system is equal to zero.

MODULE 11 - ANSWERS TO THE REVIEW QUESTIONS

1. A mass/spring system exhibits simple harmonic motion. In simple harmonic motion, the period is independent of the amplitude. Thus, the period will still be 3.0 sec.

2. A force that opposes the displacement.

3. The amplitude of the motion is the maximum displacement the object reaches from equilibrium. We learned Experiment 11.1 that the maximum displacement is the original distance that the object was pulled from equilibrium. Thus, the amplitude is 34.1 cm.

4. The restoring force in the system must be linearly proportional to and opposite of the displacement from equilibrium.

5. At the equilibrium position, because it has converted all of its energy into kinetic energy at that point.

6. At the amplitudes, because Hooke's Law tells us that the maximum force occurs at the maximum displacement from equilibrium.

7. At the amplitudes, because Equation (11.12) tells us that potential energy gets bigger the farther the object is from equilibrium.

8. The spring will compress farther the second time, because the box has more energy, so it can do more work in compressing the spring.

9. The picture on the left, because a pendulum exhibits simple harmonic motion only when the angle of displacement is small.

10. Equation (11.20) tells us that the period depends inversely on the acceleration due to gravity. Since the acceleration due to gravity is less on the moon, the period will be larger on the moon.

MODULE 12 - ANSWERS TO THE REVIEW QUESTIONS

1. In a transverse wave, the propagation is perpendicular to the oscillation. In a longitudinal wave, propagation is parallel to oscillation. Light is a transverse wave, whereas sound is a longitudinal wave.

2. Pitch is determined by frequency and volume by amplitude. Thus, the first wave has the largest frequency and the second has the largest amplitude.

3. The more dense the medium, the faster sound waves travel. Thus, substance A has the highest density.

4. There are at least four differences between sound and light waves:

   1. Sound waves are longitudinal whereas light waves are transverse.
   2. Sound waves travel slower than light waves.
   3. Sound waves are made up of air. We don't know what light waves are made of.
   4. Sound waves travel faster in denser mediums, whereas light waves travel slower in denser mediums.

5. The particle/wave duality of light means that light can be thought of as both a particle and a wave.

6. Wavelength and frequency are inversely proportional. Thus, wave B has the highest frequency.

7. Virtual images are formed by the extrapolation of light rays, whereas real images are formed when light rays intersect.

8. When light hits a transparent object, part of the light goes through the object, but the rest is reflected. You see through the window with the light that travels through it and you see your reflection with the light that is reflected.

9. The larger the index of refraction, the slower light travels. Thus, light will travel fastest in substance A.

10. The ciliary muscle of the eye changes the shape of the lens. This changes its focal point, canceling out the effect caused by the object moving.

MODULE 13 - ANSWERS TO THE REVIEW QUESTIONS

1. The proton is positively charged, the neutron has no charge, and the electron has a negative charge.

2. Despite the fact that all matter contains electrical charges, most matter has an equal number of both positive and negative charges. Thus, the negatives cancel the positives and the matter has no net electrical charge.

3. In charging by conduction, you are taking charge from one object and giving it to another. This results in the object that you are charging having the same type of charge as the object that you used. In charging by induction, you are not taking charge from the electrically charged object. Instead, you are using that charge to force the same type of charge in the other object to leave. The net result is that the object you are charging has the opposite charge as the object you used.

4. The only way to charge something so that it has the same charge as the object you are using is to charge by conduction.

5. Conductors allow charge to move freely within them, insulators do not. Metals are generally conductors while non-metals are generally not.

6. The arrows point out the direction that a positive charge would move if placed in the field. A negative charge moves opposite of the directions pointed out by the arrows.

7. The density of the electrical field lines is proportional to the electrostatic force that a charge placed in the field experiences.

8. Since like charges repel, the charges will begin to move away from each other. As they move away form each other, the distance between them increases. This means that the electrostatic force between them will decrease.

9. In electrostatics, charges exert a mutual force. That means whatever force one exerts, the other exerts an identical force. Thus, charge 2 exerts a 1,001 Newtons force on charge 1.

10. The distance between charged objects is in the denominator of Equation (13.1). This means that as the distance increases, the force decreases. Also, the distance is squared. Thus, if the distance increases by a factor of 2, the denominator in Equation (13.1) increases by a factor of 2^2, or 4. Thus, the force decreases by a factor of 4.

MODULE 14 - ANSWERS TO THE REVIEW QUESTIONS

1. Electrical potential and potential energy are not the same. They have different physical units and describe different physical phenomena.

2. The electrical potential deals only with the stationary charge. Thus, the electrical potential is negative. The potential energy is calculated by multiplying the potential (which is negative) times the freely moving particle's charge (which is positive). Thus, the potential energy is also negative.

3. The electrical potential deals only with the stationary charge. Thus, the electrical potential is positive. The potential energy is calculated by multiplying the potential (which is positive) times the freely moving particle's charge (which is negative). Thus, the potential energy is negative.

4. The unit Volt always refers to electrical potential.

5. In this case, the electrical potential is positive, but the potential energy is negative. When the particle moves closer to the stationary charge, the value for "r" in Equation (14.1) is reduced. This makes the magnitude of the electrical potential, and therefore the potential energy, greater. However, since the potential energy is negative, a larger magnitude results in a smaller number. Thus, the potential energy decreases.

6. In this case, the electrical potential is positive, as is the potential energy. When the particle moves closer to the stationary charge, the value for "r" in Equation (14.1) is reduced. This makes the magnitude of the electrical potential, and therefore the potential energy, greater. Since the potential energy is positive, a larger magnitude results in a larger number. Thus, the potential energy increases.

7. Remember, the electrical potential depends only on the stationary charge. Thus, if *any* particle feels a positive electrical potential, the stationary charge is positive.

8. According to our convention, when a particle moves from the negative plate to the positive plate, its change in electrical potential is positive. To get the potential energy, however, we need to multiply by the particle's charge. Since the charge is negative, then the change in potential energy is negative. Thus, the potential energy decreases.

9. The electron beam is deflected towards the positive plate and away from the negative plate.

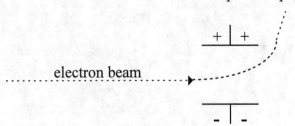

10. The electron beam causes the screen to light up. Since a dot appears in the center of the screen, we can assume that the beam is hitting the center of the screen instead of sweeping back and forth. The only reason that this could be happening is that <u>for a brief moment, the electron beam is on but the controlling capacitors are not</u>. Therefore, the beam is hitting the screen, but the television cannot sweep it yet. When the picture appears, you know that the beam is sweeping and therefore the controlling capacitors must be on.

MODULE 15 - ANSWERS TO THE REVIEW QUESTIONS

1. Conventional current is current that runs from the positive side of the battery to the negative side. Although we now know that this is physically not what happens, scientists have used it since electricity was first studied, so we continue to use it to be consistent with all of the circuit drawings that have been done in the past.

2. Electrical current is the flow of electrons through a conductor. It is defined as the amount of charge that passes through a conductor each second. As a result, the units are Coulombs per second, which we also call Amperes.

3. Resistance is the friction that occurs between electrons and the atoms in the conductor through which they are flowing. This friction removes kinetic energy from the electrons and converts it into other forms, typically heat and light.

4. All conductors, including wires, have resistance because all matter is made of atoms, and resistance is the friction that occurs between electrons and the atoms in the conductor.

5. A switch is a point on the circuit that can pull away, breaking contact with the rest of the circuit or be pushed in place, making contact with the circuit. When the switch is pulled away from the circuit, there is no longer a complete path for the current to travel from one end of the battery to the other. As a result, current ceases to flow and the light bulb turns off. When the switch is pushed back into contact with the circuit, the electrons have a complete path to travel again, so current begins to flow. This turns the light on.

6. In a series circuit, all current has to travel through every component. In a parallel circuit, the current has a choice. Some of the current travels through some of the components, and some travels through others.

7. The only way some of the lights could be lit while others are not is if the current can choose which path to take. Since the current can choose, this is a parallel circuit.

8. Fuses limit the amount of current that a circuit can draw. This is a safety precaution. A fuse is made of a heat-sensitive conductor. If too much current starts flowing, the metal gets too hot and melts. Once the metal melts, the fuse acts like an open switch, and current ceases to flow.

9. A fuse can only melt once. After that, you must replace it with a new one. A circuit breaker can be reset over and over again.

10. When several resistors are put together in a circuit, they can be reduced to a single resistor. The circuit behaves just as if that one, "effective" resistor was the only thing in the circuit.

MODULE 16 - ANSWERS TO THE REVIEW QUESTIONS

1. They repel each other.

2. Magnetic field lines point in the direction that the north pole of a magnet would point if it were placed in the field.

3. Magnetic field lines leave the north poles and go to the nearest south poles. Thus, between the magnets, we have straight lines going from one magnet's north pole to the other magnet's south pole. Remember, however, that there are magnetic field lines coming out the other side of the north poles as well. They must travel around to the magnet's south pole.

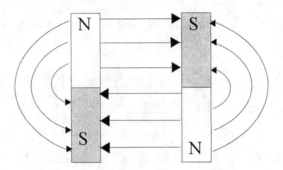

4. A dipole has both sides. For electricity, this means a positive charge on one side and a negative on the other. For magnetism, this means a north pole on one side and a south pole on the other. While electrical monopoles can exist, magnetic monopoles cannot.

5. Magnetism is produced by charged particles in motion.

6. diamagnetic, paramagnetic, and ferromagnetic

7. The electrons that orbit the nucleus of any atom each produce a magnetic field. In some atoms, the electrons are arranged so that these magnetic fields cancel. Substances made up of these atoms are called **diamagnetic** substances. If, however, the electrons in an atom do not cancel out each other's magnetic fields, then that atom is magnetic. A substance made of such atoms is either paramagnetic or ferromagnetic. If the atoms are completely misaligned so that their magnetic fields cancel out, then the substance is **paramagnetic**. If, instead, the atoms align themselves in little clusters that each behave as a tiny magnet, then the substance is **ferromagnetic**.

8. If the electrons in an atom are arranged so that their individual magnetic fields cancel, the atom is not magnetic. If the electrons' magnetic fields do not cancel, then the atom is magnetic.

9. No. A ferromagnetic substance cannot be a magnet unless its magnetic domains are aligned.

10. Only diamagnetic substances do not react to a magnet.

11. If the substance is attracted even when the magnet is weak, the substance must be <u>ferromagnetic</u>.

12. A diamagnetic substance would not at all be attracted, while a ferromagnetic substance would be strongly attracted. This, then, must be a <u>paramagnetic</u> substance.

13. Take your left hand and point its thumb in the direction of the current:

Your fingers curl in the direction of the magnetic field lines.

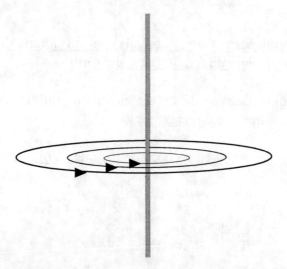

14. <u>When a conductor is exposed to a changing magnetic field, it produces a current. The direction of the magnetic field's change determines the direction of the current</u>.

15. All power plants with the exception of solar power plants use a coil turning in a magnet to generate electricity:

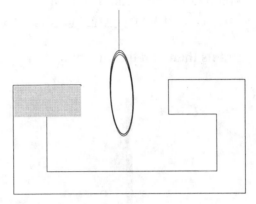

16. <u>In a hydroelectric plant, the kinetic energy of the water as it flows down the river or waterfall is used to turn a waterwheel. Thus, the kinetic energy of the water is converted into the kinetic energy of the waterwheel. The kinetic energy of the waterwheel is then converted into the kinetic energy of the turning coil. Finally, the kinetic energy of the turning coil is then converted into the kinetic energy of the electrons in the coil of wires, making electrical current.</u>

17. <u>The potential energy contained in the chemical bonds which make up wood is converted to the kinetic energy of the heat that is produced when the wood burns. That kinetic energy is then used to increase the motion of the water molecules until they break away from their liquid phase and become gas. The kinetic energy of the steam is turned into the kinetic energy of the turbines, which is turned into the kinetic energy of the coil of wire, which is finally turned into the kinetic energy of the electrons in the current.</u>

18. <u>In a direct current circuit, current always travels one way around the circuit. In an alternating current circuit, the current changes directions back and forth.</u>

19. A battery's positive and negative side are the same; thus, the current always flows the same way. This means the circuit must be <u>direct current</u>.

20. <u>No. The maximum Voltage is greater, because the rated Voltage is only the effective value.</u>

Solutions To The

Practice Problems

MODULE 1 - ANSWERS TO THE PRACTICE PROBLEMS

1. $\dfrac{1.2 \text{ mL}}{1} \times \dfrac{0.001 \text{ L}}{1 \text{ mL}} = \underline{0.0012 \text{ L}}$. Since this is a conversion, we must end with the same number of significant figures that were in our original measurement.

2. $\dfrac{34.5 \text{ km}}{1} \times \dfrac{1{,}000 \text{ m}}{1 \text{ km}} = 34{,}500 \text{ m} = \underline{3.45 \times 10^4 \text{ m}}$. Since our original measurement had only three significant figures in it, the best way I could write the answer with the same number of significant figures was to put it in scientific notation.

3. This is a two-step conversion, since we know of no relationship between km and cm. Thus, we must first convert km to m and then convert m to cm. We'll do this on one line:

$$\dfrac{0.045 \text{ km}}{1} \times \dfrac{1{,}000 \text{ m}}{1 \text{ km}} \times \dfrac{1 \text{ cm}}{0.01 \text{ m}} = 4500 \text{ cm} = \underline{4.5 \times 10^3 \text{ cm}}$$

Once again, I used scientific notation as the best way to write the answer with the proper number of significant figures.

4. This is another two-step conversion. We must convert mL to L and then L to kL since there is no direct relationship between mL and kL:

$$\dfrac{34.6 \text{ mL}}{1} \times \dfrac{0.001 \text{ L}}{1 \text{ mL}} \times \dfrac{1 \text{ kL}}{1{,}000 \text{ L}} = \underline{0.0000346 \text{ kL}}$$

5. The volume of a box is length times width times height:

$$V = 2.3 \text{ m} \cdot 4.2 \text{ m} \cdot 3.5 \text{ m} = 34 \text{ m}^3$$

Since we are multiplying, we must keep the lowest number of significant figures. All of the numbers have 2 significant figures, so there should be 2 in the final answer. Unfortunately, the mean guy who wrote the problem wants the answer in cubic centimeters (cc's), so we still have to make a conversion:

$$\dfrac{34 \text{ m}^3}{1} \times \left(\dfrac{1 \text{ cm}}{0.01 \text{ m}}\right)^3 = \dfrac{34 \text{ m}^3}{1} \times \dfrac{1 \text{ cm}^3}{0.000001 \text{ m}^3} = \underline{3.4 \times 10^7 \text{ cm}^3}$$

6. This is a simple problem if you recognize that a cubic centimeter, or cc, is the same thing as a mL:

$$\dfrac{34.5 \text{ mL}}{1} \times \dfrac{0.001 \text{ L}}{1 \text{ mL}} = \underline{0.0345 \text{ L}}$$

7. a. This is a big number, so the exponent needs to be positive. To get the decimal next to the first digit, we have to move it two places, thus $\underline{1.2345 \times 10^2}$.

   b. This is a small number, so the exponent has to be negative. The decimal point needs to be moved 4 places to get to the right of the first digit, so $\underline{3.040 \times 10^{-4}}$.

   c. This is a big number, so the exponent is positive, and the decimal must be moved 6 places to get it next to the first digit, so $\underline{6.1 \times 10^6}$.

   d. This number is small, so the exponent is negative, and the decimal need only be moved one place, so $\underline{1.234 \times 10^{-1}}$.

8. a. A positive exponent tells us to make this a big number by moving the decimal 3 places, so $\underline{6540}$.

   b. A negative exponent means to make the number small by moving the decimal 3 places, so $\underline{0.003450}$.

   c. A positive exponent tells us to make this a big number by moving the decimal 7 places, so $\underline{35600000}$.

   d. A negative exponent tells us to make the number small by moving the decimal 7 places, so $\underline{0.0000004050}$.

9. We need to use Equation (1.1) here, but we must do two things. First, we have to get the units to agree. The density is given in grams per mL, but the volume is in L. First, then, we have to convert the volume to mL:

$$\frac{3.45 \, \cancel{L}}{1} \times \frac{1 \, mL}{0.001 \, \cancel{L}} = 3.45 \times 10^3 \, mL$$

Now that the units work out, we must also rearrange Equation (1.1) to solve for mass:

$$m = \rho \cdot V$$
$$m = 11.4 \, \frac{g}{\cancel{mL}} \cdot 3.45 \times 10^3 \, \cancel{mL}$$
$$\underline{m = 3.93 \times 10^4 \, g}$$

10. This problem is like #9, but we have to convert kg to g and rearrange equation 1.1 to solve for volume:

$$\frac{45.6 \, \cancel{kg}}{1} \times \frac{1,000 \, g}{1 \, \cancel{kg}} = 4.56 \times 10^4 \, g$$

$$V = \frac{m}{\rho}$$

$$V = \frac{4.56 \times 10^4 \text{ g}}{19.3 \frac{\text{g}}{\text{cc}}}$$

$$V = \underline{2.36 \times 10^3 \text{ cc}}$$

Since the problem did not request any specific units, we can simply leave our answer in cc's.

MODULE 2 - ANSWERS TO THE PRACTICE PROBLEMS

1. The driver traveled 35.4 km one way and 13.2 km back. Thus, the total distance traveled is just 35.4 km + 13.2 km = <u>48.6 km</u>. When determining displacement, however, you must define directions. I will say that motion towards the delivery site is positive and motion towards the business is negative. Thus, his total displacement was 35.4 + -13.2 = 22.2 km. Since positive means towards the delivery site, we could say that the driver is <u>22.2 km from the business, in the direction of the delivery site</u>.

2. Average speed is calculated by taking the change in distance and dividing by the change in time. The total distance traveled was 48.6 km and the total time was 21.1 min + 7.5 min = 28.6 min. Thus, the total speed is:

$$s = \frac{48.6 \text{ km}}{28.6 \text{ min}} = 1.70 \frac{\text{km}}{\text{min}}$$

Now you might wonder why I didn't convert to meters and seconds. I could have, but there is no reason to. Usually, we only convert units if the problem asks for specific units or if the units are not consistent. This problem did not request specific units, and there was no need to worry about units being consistent, because nothing had to cancel out in the math. As a result, I just decided not to convert at all.

To determine the velocity, we need to take the change in *displacement* and divide it by the change in time. The change in displacement is +22.2 km, and the change in time is the same as before. Thus, the average velocity is:

$$v = \frac{22.2 \text{ km}}{28.6 \text{ min}} = 0.776 \frac{\text{km}}{\text{min}}$$

Since, according to my original definition of direction, positive motion means towards the delivery site, I can say <u>the velocity was 0.776 km/min in the direction of the delivery site, while the speed was 1.70 km/min</u>.

3. This problem gives you enough information to determine displacement; it gives you the average velocity, and it asks you for the time. These three quantities are related by Equation (2.1), so that's what we'll use. To determine displacement, we need to define direction. I will say that motion away from the take off site is positive and motion back towards it is negative. Thus, the displacement is 672.1 km + -321.9 km = 350.2 km. The problem is that we can't keep this unit. The velocity is given in m/sec, so I need to convert km into m:

$$\frac{350.2 \cancel{\text{km}}}{1} \times \frac{1,000 \text{ m}}{1 \cancel{\text{km}}} = 3.502 \times 10^5 \text{ m}$$

Now we can use Equation (2.1):

$$v = \frac{\Delta x}{\Delta t}$$

$$42 \frac{m}{sec} = \frac{3.502 \times 10^5 \, m}{\Delta t}$$

$$\Delta t = \frac{3.502 \times 10^5 \, m}{42 \frac{m}{sec}} = 8.3 \times 10^3 \, sec$$

This means that the flight took 8.3×10^3 seconds, or 2.3 hours.

4. The graph starts out with a positive slope, indicating that velocity is positive. The slope turns negative just after 6 seconds. This means that the car changed directions just after 6.0 seconds.

5. The car moves fastest where the slope is steepest. The curve is steepest from about 5.5 seconds to about 6.0 seconds.

6. The curve slopes steeper at 10.5 seconds than it does at 4.0 seconds. This means that the car moves faster at 10.5 seconds. Remember, the sign of the slope only tells us direction, so negative steep slopes represent velocities that are just as fast as positive steep slopes.

7. This problem gives you velocity and acceleration, and then it asks you to calculate time. Equation (2.3) relates these quantities, but we need to be careful. The train starts out with a velocity of 20.1 m/sec. Once it stops, the velocity will be zero. Thus, Δv = 0 m/sec - 20.1 m/sec = -20.1 m/sec. This is what we use in Equation (2.3):

$$a = \frac{\Delta v}{\Delta t}$$

$$-0.0500 \frac{m}{sec^2} = \frac{-20.1 \frac{m}{sec}}{\Delta t}$$

$$\Delta t = \frac{-20.1 \frac{m}{sec}}{-0.0500 \frac{m}{sec^2}} = 402 \, sec$$

If you had not used the proper definition of Δv, then you would have gotten a negative time. Hopefully that would have alerted you to the problem, since time can never be negative! Thus, the train takes 402 seconds to stop.

8. The runner slows down whenever the acceleration and velocity have opposite signs. At about 8.0 seconds, the velocity is still positive, but the slope (acceleration) is negative. This continues until 15.0 seconds. At that point, the velocity becomes negative, and the slope (acceleration) is also negative. That means that the runner is speeding up again. However, after 19.0 seconds, the velocity is still negative, but the slope (acceleration) is positive. This means that the runner is slowing down during this time interval as well. The runner, then, is slowing down from 8.0 seconds to 15.0 seconds and again from 19.0 seconds to 20.0 seconds.

9. At 6.0 seconds, the curve is flat. This means that the acceleration is zero.

10. From 0.0 to 3.0 seconds, the curve looks like a straight line. At 0.0 seconds, velocity is 0.0 and at 3.0 seconds, the velocity looks to be about 6.0 m/sec. The slope, then, is:

$$\text{slope} = \frac{6.0\ \frac{m}{\sec} - 0.0\ \frac{m}{s}}{3.0\ \sec - 0.0\ \sec} = 2.0\ \frac{m}{\sec^2}$$

This means that at any time from 0.0 to 3.0 seconds, the slope is 2.0 m/sec². At 2.2 seconds, then, the acceleration is 2.0 m/sec².

MODULE 3 - ANSWERS TO THE PRACTICE PROBLEMS

1. Since the jogger is moving at a constant velocity, his or her acceleration is zero. Since the velocity is constant, we also know that we can use it as either the initial or final velocity, or both. So, we have initial and/or final velocity, acceleration, and time, and we want to determine distance. Equation (3.19) will work:

$$x = v_o t + \frac{1}{2}at^2$$

Before we use the equation, however, we need to make our units work. Right now, time is in minutes while velocity is in m/sec. To fix this, we will convert minutes to seconds:

$$\frac{25 \text{ min}}{1} \times \frac{60 \text{ sec}}{1 \text{ min}} = 1.5 \times 10^3 \text{ sec}$$

Now that our units agree, we can solve the equation:

$$x = (5.0 \frac{m}{sec}) \cdot (1.5 \times 10^3 \text{ sec}) + \frac{1}{2} \cdot 0 \cdot (1.5 \times 10^3 \text{ sec})$$

$$x = \underline{7.5 \times 10^3 \text{ m}}$$

2. In this problem, we are given initial and final velocity, as well as time. We need to determine acceleration. Equation (3.6) relates these quantities:

$$v = v_0 + at$$

Before we use the equation, however, we need to get our units worked out. The velocity is in miles per hour while the time is in seconds. Because we usually talk about cars in miles per hour, I will convert the time into hours:

$$\frac{7.9 \text{ sec}}{1} \times \frac{1 \text{ hr}}{3600 \text{ sec}} = 0.0022 \text{ hr}$$

Now we can use the equation:

$$60.0 \frac{mi}{hr} = 0.0 + a \cdot (0.0022 \text{ hr})$$

$$a = \frac{60.0 \frac{mi}{hr}}{0.0022 \text{ hr}} = 2.7 \times 10^4 \frac{mi}{hr^2}$$

So the BMW has an acceleration of $\underline{2.7 \times 10^4 \text{ mi/hr}^2}$.

3. Including what we just calculated in problem #2, we have the initial and final velocity of the BMW, as well as acceleration and time. With all of this information I can use either Equation (3.15) or (3.19) to solve the problem. I will use Equation (3.19):

$$x = (0) \cdot (0.0022 \text{ hr}) + \frac{1}{2} \cdot (2.7 \times 10^4 \frac{\text{mi}}{\text{hr}^2}) \cdot (0.0022 \text{hr})^2$$

$$x = 0.065 \text{ mi}$$

The BMW, then, travels 0.065 miles while it is accelerating.

4. This problem gives us initial velocity (25 m/s), final velocity (0), and displacement (0.103 km). We need to calculate acceleration, so we will use Equation (3.15);

$$v^2 = v_0^2 + 2ax$$

Before we use the equation, however, we need to make our units consistent:

$$\frac{0.103 \text{ km}}{1} \times \frac{1000 \text{ m}}{1 \text{ km}} = 103 \text{ m}$$

We can now use the equation:

$$(0)^2 = (25 \frac{\text{m}}{\text{sec}})^2 + 2a \cdot (103 \text{ m})$$

$$a = \frac{-(25 \frac{\text{m}}{\text{sec}})^2}{2 \cdot 103 \text{ m}} = -3.0 \frac{\text{m}}{\text{sec}^2}$$

The acceleration is negative because it opposes the positive velocity. So the deceleration is -3.0 m/sec^2.

5. This problem gives us displacement (2.0 x 10^3 yds), initial velocity (0), and acceleration (9.7 x 10^4 ft/sec^2), and we need to determine the final velocity. Equation (3.15) will be our tool:

$$v^2 = v_0^2 + 2ax$$

We can't use this equation until our units are consistent, however. To do this, I will convert yards into feet:

$$\frac{2.0 \times 10^3 \text{ yds}}{1} \times \frac{3 \text{ ft}}{1 \text{ yd}} = 6.0 \times 10^3 \text{ ft}$$

Now we can use the equation:

$$v^2 = (0)^2 + 2 \cdot (9.7 \times 10^4 \frac{\text{ft}}{\text{sec}^2}) \cdot (6.0 \times 10^3 \text{ ft})$$

$$v = 3.4 \times 10^4 \frac{\text{ft}}{\text{sec}}$$

The rocket traveled with a velocity of 3.4×10^4 ft/sec after it is done accelerating. This is just about the necessary velocity to escape the pull of earth's gravity and go into space.

6. The ride starts from rest, so the initial velocity is 0. The displacement is 30 m, and, since the ride is in free fall, the acceleration is 9.8 m/sec². Notice that since I left the displacement positive, this means that downward motion is positive. That's why the acceleration is positive as well. We have initial velocity, acceleration, and distance, and we want to know final velocity. We will therefore use Equation (3.15). Since our units are all consistent, we can just plug things directly into the equation:

$$v^2 = v_0^2 + 2ax$$

$$v^2 = (0)^2 + 2 \cdot (9.8 \frac{\text{m}}{\text{sec}^2}) \cdot (30 \text{ m})$$

$$v = 24 \frac{\text{m}}{\text{sec}}$$

So the velocity of the ride at the bottom of the fall is 24 m/sec down, which is about 55 mph!

7. In this problem, we are given the time it takes a rock to fall. If we can calculate its displacement in that time, we will be able to determine the height of the cliff. Since the rock is in free fall, we know its acceleration is 9.8 m/sec². We also know that since it was dropped, its initial velocity is zero. I went ahead and left the acceleration positive, indicating that down is the positive direction. If we have initial velocity, acceleration, and time, and we want to determine displacement, Equation (3.19) is the one to use:

$$x = v_0 t + \frac{1}{2} at^2$$

$$x = (0) \cdot (3.5 \text{ sec}) + \frac{1}{2} \cdot (9.8 \frac{\text{m}}{\text{sec}^2}) \cdot (3.5 \text{ sec})^2 = 6.0 \times 10^1 \text{ m}$$

The cliff, then, is 6.0×10^1 m high.

8. This problem gives us time (5.0 sec) and, since we know that without the parachute the man is in free fall, the acceleration (9.8 m/sec^2). We also know that the initial velocity is zero, because parachutists drop out of planes. Since I left the acceleration positive, it means that down is the positive direction. We know time, initial velocity, and acceleration, and we want to determine final velocity, so Equation (3.6) is our tool. All of our units work out, so we can just plug into the equation:

$$\mathbf{v} = \mathbf{v}_o + \mathbf{a}t$$

$$v = 0 + (9.8 \frac{m}{sec^2}) \cdot (5.0 \ \cancel{sec}) = 49 \frac{m}{sec}$$

The final velocity is <u>49 m/sec</u>, which is nearly 100 mph!

9. As we learned in this module, the ball will reach its maximum height when its velocity is zero. The initial velocity is 3.9 m/sec. If we keep that positive, it tells us that upward motion is positive. In that case, the acceleration due to gravity is -9.8 m/sec^2. So, we know final velocity, initial velocity, and acceleration, and we want to determine distance. Equation (3.15) will help us with that. Since all of our units are consistent, we can plug our numbers right into the equation:

$$v^2 = v_o^2 + 2ax$$

$$(0)^2 = (3.9 \frac{m}{sec})^2 + 2 \cdot (-9.8 \frac{m}{sec^2}) \cdot (x)$$

$$x = \frac{-(3.9 \frac{m}{\cancel{sec}})^2}{-19.6 \frac{m}{\cancel{sec^2}}} = 0.78 \ m$$

The ball reaches a maximum height of <u>0.78 m</u>.

10. In this problem, we are given the time (4.2 seconds) and, since we know that the ball is in free fall, we also know the acceleration. Because there is some upward motion in this problem, I will go ahead and define upward motion as positive. This means that the acceleration is -32 ft/sec^2. I used the English units here because the height of the ladder was given in feet. Since that height is 12 ft, we know that when the ball hits the ground, its displacement is 12 ft down from where it began. Since down is negative, that tells us that the displacement is -12 ft. To relate these quantities to initial velocity, we will have to use Equation (3.19). Since I used the English units for acceleration, all of our units agree, and we can simply plug them into the equation:

$$x = v_o t + \frac{1}{2}at^2$$

$$-12 \text{ ft} = v_o \cdot (4.2 \text{ sec}) + \frac{1}{2} \cdot (-32\frac{\text{ft}}{\text{sec}^2}) \cdot (4.2 \text{ sec})^2$$

$$v_o = \frac{-12 \text{ ft} + (16\frac{\text{ft}}{\text{sec}^2}) \cdot (4.2 \text{ sec})^2}{4.2 \text{ sec}} = 64 \frac{\text{ft}}{\text{sec}}$$

The ball was thrown up (that's why the velocity is positive) at 64 ft/sec.

MODULE 4 - ANSWERS TO THE PRACTICE PROBLEMS

1. To add these two vectors, we must move the second so that its tail is at the head of the first:

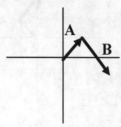

The vector that starts at the tail of the first and points to the head of the second represents the sum:

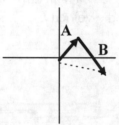

The dashed arrow, then, represents **A** + **B**.

2. To subtract **B** from **A**, we must first take the negative of **B**:

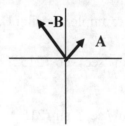

Then we add **-B** to **A**:

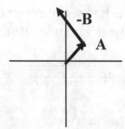

The vector representing the difference is then drawn by starting at the tail of the first vector and pointing to the head of the second:

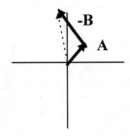

The dashed arrow, then, represents **A - B**.

3. This is just like Problem #1, because when an airplane is flying, its true velocity will be the vector sum of its velocity and the wind's velocity:

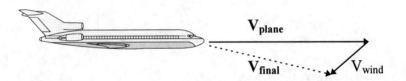

The dashed arrow is the plane's true velocity.

4. This problem is a straightforward example of using Equations (4.6) and (4.7):

$$A_x = (35.1 \text{ m/sec}) \cdot \cos(215°) = \underline{-28.8 \text{ m/sec}}$$

$$A_y = (35.1 \text{ m/sec}) \cdot \sin(215°) = \underline{-20.1 \text{ m/sec}}$$

5. In this problem, we are given the x- and y-components of a vector and are asked to calculate its magnitude and direction. To get the magnitude, we use Equation (4.2):

$$\text{Magnitude} = \sqrt{V_x^2 + V_y^2} = \sqrt{(17 \text{ m/sec})^2 + (-11 \text{ m/sec})^2} = \underline{2.0 \times 10^1 \text{ m/sec}}$$

To get the angle, we start with Equation (4.3):

$$\theta = \tan^{-1}(\frac{V_y}{V_x}) = \tan^{-1}(\frac{-11\frac{m}{\sec}}{17\frac{m}{\sec}}) = \tan^{-1}(-0.647) = -33°$$

We aren't necessarily finished yet, however. We have to determine which region of the Cartesian coordinate system that the vector is in. Since the x-component is positive and the y-component is negative, the vector is to the right and below the origin, which means that the vector is in region IV. According to our rules, we add 360.0 degrees to the result of Equation (4.3) when the vector is in region IV, so 327° is the proper angle. Thus, the vector has magnitude of 2.0×10^1 m/sec and direction of 327°.

6. The first step in adding vectors analytically is to break both vectors down into their components:

$$A_x = (31.1 \text{ m}) \cdot \cos(160.0°) = -29.2 \text{ m}$$

$$A_y = (31.1 \text{ m}) \cdot \sin(160.0°) = 10.6 \text{ m}$$

$$B_x = (11.4 \text{ m}) \cdot \cos(290.0°) = 3.90 \text{ m}$$

$$B_y = (11.4 \text{ m}) \cdot \sin(290.0°) = -10.7 \text{ m}$$

Now that we have the individual components, we can add them like numbers.

$$C_x = A_x + B_x = -29.2 \text{ m} + 3.90 \text{ m} = -25.3 \text{ m}$$

$$C_y = A_y + B_y = 10.6 \text{ m} + -10.7 \text{ m} = -0.1 \text{ m}$$

Now that we have the components to our answer, we can use Equations (4.2) and (4.3) to give us the magnitude and direction of the answer. To get the magnitude of this vector, we use Equation (4.2):

$$\text{Magnitude} = \sqrt{C_x^2 + C_y^2} = \sqrt{(-25.3 \text{ m})^2 + (-0.1 \text{ m})^2} = 25.3 \text{ m}$$

To find the direction of the vector, we use Equation (4.3):

$$\theta = \tan^{-1}(\frac{C_y}{C_x}) = \tan^{-1}(\frac{-0.1 \text{ m}}{-25.3 \text{ m}}) = 0.2°$$

Since the x and y-components are both negative, the vector is in the region III of the Cartesian coordinate plane, so we add 180.0 degrees to the result of Equation (4.3). Thus, the sum of vectors **A** and **B** has a magnitude of 25.3 m at a direction of 180.2°.

7. In this two-dimensional problem, we are given two displacement vectors followed by the boy scout, and we are asked to come up with his final displacement vector. The final vector must be the sum of the two. In the end, then, we simply have to add these two vectors. The first step is to break both vectors down into their components:

$$A_x = (1.2 \text{ miles}) \cdot \cos(150.0°) = -1.0 \text{ miles}$$

$$A_y = (1.2 \text{ miles}) \cdot \sin(150.0°) = 0.60 \text{ miles}$$

$$B_x = (3.2 \text{ miles}) \cdot \cos(250.0°) = -1.1 \text{ miles}$$

$$B_y = (3.2 \text{ miles}) \cdot \sin(250.0°) = -3.0 \text{ miles}$$

Now that we have the individual components, we can add them like numbers.

$$C_x = A_x + B_x = -1.0 \text{ miles} + -1.1 \text{ miles} = -2.1 \text{ miles}$$

$$C_y = A_y + B_y = 0.60 \text{ miles} + -3.0 \text{ miles} = -2.4 \text{ miles}$$

Now that we have the components to our answer, we can use Equations (4.2) and (4.3) to give us the magnitude and direction of the answer. To get the magnitude of this vector, we use Equation (4.2):

$$\text{Magnitude} = \sqrt{C_x^2 + C_y^2} = \sqrt{(-2.1 \text{ miles})^2 + (-2.4 \text{ miles})^2} = 3.2 \text{ miles}$$

To find the direction of the vector, we use Equation (4.3):

$$\theta = \tan^{-1}\left(\frac{C_y}{C_x}\right) = \tan^{-1}\left(\frac{-2.4 \text{ miles}}{-2.1 \text{ miles}}\right) = 49°$$

Since the both components of vector **C** are negative, the vector is in region III of the Cartesian coordinate plane. This means that we need to add 180.0° to the result of Equation (4.3) in order to get the angle defined properly. Thus, the proper angle is 49° + 180.0° = 229°. So the scout's final displacement is 3.2 miles at a direction of 229°.

8. Since the boat will be carried by the river's current, the actual velocity of the boat will be the vector sum of the boat and the current. To do this, we first split them up into their components:

$$V_{boat_x} = (25 \text{ mph}) \cdot \cos(30.0°) = 22 \text{ mph}$$

$$V_{boat_y} = (25 \text{ mph}) \cdot \sin(30.0°) = 13 \text{ mph}$$

$$V_{current_x} = (5.1 \text{ mph}) \cdot \cos(270.0°) = 0$$

$$V_{current_y} = (5.1 \text{ mph}) \cdot \sin(270.0°) = -5.1 \text{ mph}$$

Now we add the components together to get the components of the final velocity:

$$V_{final_x} = 22 \text{ mph} + 0 \text{ mph} = 22 \text{ mph}$$

$$V_{final_y} = 13 \text{ mph} + -5.1 \text{ mph} = 8 \text{ mph}$$

With these components, we can get the magnitude and direction of the final velocity:

$$\text{Magnitude} = \sqrt{V_{final_x}^2 + V_{final_y}^2} = \sqrt{(22 \text{ mph})^2 + (8 \text{ mph})^2} = 23 \text{ mph}$$

To find the direction of the vector, we use Equation (4.3):

$$\theta = \tan^{-1}\left(\frac{V_{final_y}}{V_{final_x}}\right) = \tan^{-1}\left(\frac{8 \text{ mph}}{22 \text{ mph}}\right) = 20°$$

Since both components are positive, we are in region I and nothing has to be done to this angle. The boat's final velocity, then, is __23 mph at 20°__.

9. The final velocity of the plane will be the vector sum of the plane's velocity and the wind's velocity.

$$V_{plane_x} = (200.0 \text{ mph}) \cdot \cos(180°) = -200.0 \text{ mph}$$

$$V_{plane_y} = (200.0 \text{ mph}) \cdot \sin(180°) = 0$$

$$V_{wind_x} = (15 \text{ mph}) \cdot \cos(315°) = 11 \text{ mph}$$

$$V_{wind_y} = (15 \text{ mph}) \cdot \sin(315°) = -11 \text{ mph}$$

Now we add the components together to get the components of the final velocity:

$$V_{final_x} = -200.0 \text{ mph} + 11 \text{ mph} = -189 \text{ mph}$$

$$V_{final_y} = 0 \text{ mph} + -11 \text{ mph} = -11 \text{ mph}$$

With these components, we can get the magnitude and direction of the final velocity:

$$\text{Magnitude} = \sqrt{V_{final_x}^2 + V_{final_y}^2} = \sqrt{(-189 \text{ mph})^2 + (-11 \text{ mph})^2} = 189 \text{ mph}$$

To find the direction of the vector, we use Equation (4.3):

$$\theta = \tan^{-1}\left(\frac{V_{final_y}}{V_{final_x}}\right) = \tan^{-1}\left(\frac{-11 \text{ mph}}{-189 \text{ mph}}\right) = \tan^{-1}(0.0521) = 3.3°$$

Since both components are negative, the final vector is in region III of the Cartesian coordinate plane. Therefore, we must add 180 degrees to the result of Equation (4.3). The final velocity, then, is <u>189 mph at 183.3°</u>.

10. Since the boat will be carried by the river's current, the actual velocity of the boat will be the vector sum of the boat and the current. To do this, we first split them up into their components:

$$V_{boat_x} = (3.0 \text{ m/sec}) \cdot \cos(0.000°) = 3.0 \text{ m/sec}$$

$$V_{boat_y} = (3.0 \text{ m/sec}) \cdot \sin(0.000°) = 0$$

$$V_{current_x} = (1.5 \text{ m/sec}) \cdot \cos(135°) = -1.1 \text{ m/sec}$$

$$V_{current_y} = (1.5 \text{ m/sec}) \cdot \sin(135°) = 1.1 \text{ m/sec}$$

Now we add the components together to get the components of the final velocity:

$$V_{final_x} = 3.0 \text{ m/sec} + -1.1 \text{ m/sec} = 1.9 \text{ m/sec}$$

$$V_{final_y} = 0 \text{ m/sec} + 1.1 \text{ m/sec} = 1.1 \text{ m/sec}$$

With these components, we can get the magnitude and direction of the final velocity:

$$\text{Magnitude} = \sqrt{V_{final_x}^2 + V_{final_y}^2} = \sqrt{(1.9 \text{ m/sec})^2 + (1.1 \text{ m/sec})^2} = 2.2 \text{ m/sec}$$

To find the direction of the vector, we use Equation (4.3):

$$\theta = \tan^{-1}\left(\frac{V_{final_y}}{V_{final_x}}\right) = \tan^{-1}\left(\frac{1.1 \frac{m}{sec}}{1.9 \frac{m}{sec}}\right) = 3.0 \times 10^1 \text{ °}$$

Since both components are positive, we are in region I and nothing has to be done to this angle. The boat's final velocity, then, is 2.2 m/sec at 3.0×10^1 °.

MODULE 5 - ANSWERS TO THE PRACTICE PROBLEMS

1. As with all such problems, we need to look for the one-dimensional situations within this two-dimensional problem. Each leg of the journey is an independent one-dimensional problem, because the ship travels in a straight line in each case. In the first leg of the journey, the velocity is 21.1 mph, the acceleration is 0 (because the velocity is constant), and the time is 3.20 hours. Equation (3.19), then, looks like this:

$$x = v_o \cdot t + \frac{1}{2} \cdot a \cdot t^2$$

$$x = (21.1 \frac{\text{miles}}{\text{hr}}) \cdot (3.20 \text{ hr}) + \frac{1}{2} \cdot 0 \cdot (3.20 \text{hr})^2$$

$$x = 67.5 \text{ miles}$$

Is this our answer? Of course not! This was only the first leg of the journey. We did learn something, however. We learned that the first leg of the journey resulted in a displacement of 67.5 miles at an angle of 123°. That, then, is where the second leg of the journey begins. If we can figure out the displacement that results from this second leg, we should have enough information to determine the final displacement of the ship.

Well, we can analyze the second leg of the journey the same way that we analyzed the first leg.

$$x = v_o \cdot t + \frac{1}{2} \cdot a \cdot t^2$$

$$x = (18.2 \frac{\text{miles}}{\text{hr}}) \cdot (1.10 \text{ hr}) + \frac{1}{2} \cdot 0 \cdot (1.10 \text{ hr})^2$$

$$x = 20.0 \text{ miles}$$

What do we have now? Well, we know that the first leg of the journey resulted in a displacement of 67.5 miles at 123 degrees, while the second leg resulted in a displacement of 20.0 miles at 190 degrees. How can we find the total displacement? We can add them in order to get the final displacement.

To add these two displacements, we must first break the vectors down into their components. To make the notation easier, let's call the displacement vector resulting from the first leg of the journey vector **A**. We will therefore call the second vector **B**, and the final displacement vector will be known as **C**.

$$A_x = (67.5 \text{ miles}) \cdot \cos(123) = -36.8 \text{ miles}$$

$$A_y = (67.5 \text{ miles}) \cdot \sin(123) = 56.6 \text{ miles}$$

$$B_x = (20.0 \text{ miles}) \cdot \cos(190.0) = -19.7 \text{ miles}$$

$$B_y = (20.0 \text{ miles}) \cdot \sin(190.0) = -3.47 \text{ miles}$$

$$C_x = -36.8 \text{ miles} + -19.7 \text{ miles} = -56.5 \text{ miles}$$

$$C_y = 56.6 \text{ miles} + -3.47 \text{ miles} = 53.1 \text{ miles}$$

Now we can determine the magnitude and direction of the final displacement vector:

$$C = \sqrt{C_x^2 + C_y^2} = \sqrt{(-56.5 \text{ miles})^2 + (53.1 \text{ miles})^2} = 77.5 \text{ miles}$$

$$\theta = \tan^{-1}\left(\frac{53.1 \text{ miles}}{-56.5 \text{ miles}}\right) = \tan^{-1}(-.940) = -43.2°$$

Based on our rules regarding Equation (4.3), the angle for the final displacement vector is $180.0°$ + $-43.2 = 136.8°$. In the end, then, the ship's final displacement vector is <u>77.5 miles at an angle of 136.8°</u>.

2. We can solve this problem just like we solved the previous one, but we have to recognize that the units are not consistent. The velocities are given in m/sec while the time is in minutes. Thus, we need to convert minutes to seconds:

$$\frac{45 \text{ min}}{1} \times \frac{60 \text{ sec}}{1 \text{ min}} = 2.7 \times 10^3 \text{ sec}$$

Now that our units are consistent, we can use Equation (3.19) to figure out the displacement on the first leg of the journey:

$$\mathbf{x} = \mathbf{v_o} \cdot t + \frac{1}{2} \cdot \mathbf{a} \cdot t^2$$

$$\mathbf{x} = (1.2 \frac{m}{\text{sec}}) \cdot (2.7 \times 10^3 \text{ sec}) + \frac{1}{2} \cdot 0 \cdot (2.7 \times 10^3 \text{ sec})^2$$

$$x = 3.2 \times 10^3 \text{ m}$$

We can analyze the second leg of the journey in the same way. Just like we did in the first leg, however, we must convert the units on time from minutes to seconds. Once we do that, Equation (3.19) looks like this:

$$\mathbf{x} = \mathbf{v}_o \cdot t + \frac{1}{2} \cdot \mathbf{a} \cdot t^2$$

$$x = (1.2 \frac{m}{\text{sec}}) \cdot (1.9 \times 10^3) \text{ sec} + \frac{1}{2} \cdot 0 \cdot (1.9 \times 10^3 \text{ sec})^2$$

$$x = 2.3 \times 10^3 \text{ m}$$

Just like we did in the last problem, we can add the two displacements in order to get the final displacement.

To add these two displacements, we must first break the vectors down into their components. To make the notation easier, let's call the displacement vector resulting from the first leg of the journey vector **A**. We will therefore call the second vector **B**, and the final displacement vector will be known as **C**.

$$A_x = (3.2 \times 10^3 \text{ m}) \cdot \cos(45) = 2.3 \times 10^3 \text{ m}$$

$$A_y = (3.2 \times 10^3 \text{ m}) \cdot \sin(45) = 2.3 \times 10^2 \text{ m}$$

$$B_x = (2.3 \times 10^3 \text{ m}) \cdot \cos(330) = 2.0 \times 10^3 \text{ m}$$

$$B_y = (2.3 \times 10^3 \text{ m}) \cdot \sin(330) = -1.2 \times 10^3 \text{ m}$$

$$C_x = 2.3 \times 10^3 \text{ m} + 2.0 \times 10^3 \text{ m} = 4.3 \times 10^3 \text{ m}$$

$$C_y = 2.3 \times 10^3 \text{ m} + -1.2 \times 10^3 \text{ m} = 1.1 \times 10^3 \text{ m}$$

Now we can determine the magnitude and direction of the final displacement vector:

$$C = \sqrt{C_x^2 + C_y^2} = \sqrt{(4.3 \times 10^3 \text{ m})^2 + (1.1 \times 10^3 \text{ m})^2} = 4.4 \times 10^3 \text{ m}$$

$$\theta = \tan^{-1}\left(\frac{1.1 \times 10^3 \text{ m}}{4.3 \times 10^3 \text{ m}}\right) = 14°$$

Based on our rules regarding Equation (4.3), we needn't do anything to this angle to make it consistent with our definition of vector angles. In the end, then, your final displacement vector is 4.4 x 10³ m at an angle of 14°.

3. The only thing we need to calculate here is the maximum height of the projectiles. This means we are concentrating only on the y-dimension. Thus, we had better get the y-component of the initial velocity:

$$V_{o_y} = (200.0 \text{ m/sec}) \cdot \sin(30.0) = 1.00 \times 10^2 \text{ m/sec}$$

We now know the initial velocity in the y-dimension (1.00 x 10² m/sec), the acceleration (-9.8 m/sec²), and the final velocity (at the maximum height, the final velocity is 0). From these data, we need to calculate the height. Equation (3.15) will do that:

$$v^2 = v_o^2 + 2ax$$

$$0^2 = (1.00 \times 10^2 \tfrac{m}{sec})^2 + 2 \cdot (-9.8 \tfrac{m}{sec^2}) \cdot x$$

$$x = \frac{(1.00 \times 10^2 \tfrac{m}{sec})^2}{2 \cdot (9.8 \tfrac{m}{sec^2})} = 5.1 \times 10^2 \text{ m}$$

This tells us that the maximum height the projectiles reach is 5.1x10² m. Since this is significantly lower than the height of the tallest mountain, there is no chance of the projectiles hitting the target. The factory was placed there specifically to guard against ship-based attacks.

4. In solving this problem, we first have to see what dimension we are dealing with. Since the question asks about height, we are at first only interested in the y-dimension. Thus, we should get the y-component of the velocity before we start:

$$V_{o_y} = (8.2 \text{ ft/sec}) \cdot \sin(40.0°) = 5.3 \text{ ft/sec}$$

Now we have the initial velocity in the dimension of interest. In this dimension, we also know the acceleration (-32 ft/sec²), because in the y-dimension, gravity is at work. Finally, we also know the final velocity, because at a projectile's maximum height, the final velocity is zero. With these data, we are asked to calculate time. We therefore must use Equation (3.6):

$$\mathbf{v} = \mathbf{v}_0 + \mathbf{a} \cdot t$$

$$0 = 5.3 \frac{ft}{sec} + (-32 \frac{ft}{sec^2}) \cdot t$$

$$t = \frac{5.3 \frac{ft}{sec}}{32 \frac{ft}{sec^2}} = 0.16 \text{ sec}$$

Notice the signs I used. If we define the upward initial velocity as positive, then the downward acceleration must be negative. Thus, our answer tells us that it takes <u>0.16 seconds to reach maximum height</u>.

It turns out that this is all the information we need to answer the second part of the problem as well. After all, we know that when the projectile lands at the same height from which it is launched, it reaches maximum height at the halfway point of its journey. Thus, if took 0.16 seconds to reach its maximum height, it takes <u>0.32 seconds to hit the ground again</u>.

5. This is a simple application of Equation (5.9). We are given the initial speed (350 ft/sec) and the angle (35°). From those facts, we are asked to calculate the range of the projectile:

$$\text{Range} = \frac{v_0^2 \cdot \sin 2\theta}{g}$$

$$\text{Range} = \frac{(350 \frac{ft}{sec})^2 \cdot \sin(2 \cdot 35°)}{32 \frac{ft}{sec^2}} = 3.6 \times 10^3 \text{ ft}$$

So the cannonball's range is <u>3.6×10^3 ft</u>.

6. In this use of Equation (5.9), we are given the angle (45°) and the desired range (31 m). We need to determine the necessary initial speed. Thus, we just need to rearrange Equation (5.9) to solve for v_0 after we have plugged in the numbers that we know:

$$\text{Range} = \frac{v_o^2 \cdot \sin 2\theta}{g}$$

$$31 \text{ m} = \frac{v_o^2 \cdot \sin(2 \cdot 45)}{9.8 \frac{\text{m}}{\text{sec}^2}}$$

$$v_o^2 = \frac{31 \text{ m} \cdot 9.8 \frac{\text{m}}{\text{sec}^2}}{1}$$

$$v_o = 17 \frac{\text{m}}{\text{sec}}$$

So, in order to hit his teammate, the ball must be kicked with initial speed of <u>17 m/sec</u>.

7. In this application of Equation (5.9), we are given the initial speed (81 ft/sec) and the desired range (195 feet). With that information, we must calculate the angle at which the kicker should kick the ball. To do that, we will plug in our numbers and then rearrange the equation to solve for θ. Before we do this, however, notice that the units are in feet and seconds. This tells us that we need to use 32 ft/sec^2 for the acceleration due to gravity so that the units will be consistent.

$$\text{Range} = \frac{v_o^2 \cdot \sin 2\theta}{g}$$

$$195 \text{ ft} = \frac{(81 \frac{\text{ft}}{\text{sec}})^2 \cdot \sin 2\theta}{32 \frac{\text{ft}}{\text{sec}^2}}$$

$$\sin 2\theta = \frac{195 \text{ ft} \cdot 32 \frac{\text{ft}}{\text{sec}^2}}{6561 \frac{\text{ft}^2}{\text{sec}^2}}$$

$$2\theta = 72°$$

$$\theta = 36°$$

The kicker, then, must kick the ball at an angle of <u>36 degrees</u>.

8. In this problem, we want to determine the height of a hill. Thus, we are interested in the y-dimension. We can figure out the y-component of the initial velocity, and we also know the acceleration in the y-dimension. Unfortunately, to determine the distance that the ball travels in the y-dimension, we need to figure out the time it takes for the ball to hit the ground. We can do that by looking at the x-dimension.

In that dimension, we can calculate the initial velocity, and we know that the acceleration is zero. We also know that the ball traveled 94 ft in the x-dimension. With this information, we can use Equation (3.19) to determine the time that the ball was in the air. First, we need to determine the x-component of the initial velocity:

$$V_o = (75 \text{ ft / sec}) \cdot \cos(60.0) = 38 \text{ ft / sec}$$

Now we can take the data we know for the x-dimension and plug them into Equation (3.19). At that point, the only unknown will be the time, and we can solve for it:

$$x = v_o \cdot t + \frac{1}{2} \cdot a \cdot t^2$$

$$94 \text{ ft} = (38 \frac{\text{ft}}{\text{sec}}) \cdot t + \frac{1}{2} \cdot (0) \cdot t^2$$

$$t = \frac{94 \text{ ft}}{38 \frac{\text{ft}}{\text{sec}}}$$

$$t = 2.5 \text{ sec}$$

Now that we have the time, we can go back to the y-dimension and figure out the height of the hill. To do that, we need to determine the initial velocity in the y-dimension:

$$V_o = (75 \text{ ft / sec}) \cdot \sin(60.0) = 65 \text{ ft / sec}$$

We can take this initial velocity, the acceleration (-32 m/sec^2), and the time (2.5 sec) and plug them into Equation (3.19) to determine the distance that the ball falls in the y-dimension. This is equal to the height of the hill:

$$x = v_o \cdot t + \frac{1}{2} \cdot a \cdot t^2$$

$$x = (65 \frac{\text{ft}}{\text{sec}}) \cdot (2.5 \text{ sec}) + \frac{1}{2} \cdot (-32 \frac{\text{ft}}{\text{sec}^2}) \cdot (2.5 \text{ sec})^2$$

$$x = 63 \text{ ft}$$

Thus, the hill is <u>63 ft high</u>.

9. In this problem, of course, we cannot use Equation (5.9), because the ball lands below the height from which it was thrown. We must analyze each dimension, then, and figure out what to use where. We are asked to calculate the initial velocity of the ball. If we look at the initial velocity in each of the two dimensions, we will quickly see which dimension allows us to calculate that quantity:

$$V_{o_x} = V_o \cdot \cos(0.000) = V_o$$

$$V_{o_y} = V_o \cdot \sin(0.000) = 0$$

So, because the ball is thrown horizontally, the y-component of the velocity is zero. As a result, we can not calculate the initial velocity using the y-dimension. The x-dimension has a non-zero initial velocity, but we do not have enough information yet. In the x-dimension, we know the displacement (51 ft) and acceleration (0). If we could find out the time, we could use Equation (3.19) to determine the initial velocity. Unfortunately, we don't have the time yet. Let's see if we can get that from the y-dimension.

In the y-dimension, we know the initial velocity (0), the acceleration (-32 ft/sec^2), and the displacement (-2.9 ft). That's all we need to calculate the time from Equation (3.19).

$$x = v_o \cdot t + \frac{1}{2} \cdot a \cdot t^2$$

$$-2.9 \text{ ft} = (0) \cdot t + \frac{1}{2} \cdot (-32 \frac{\text{ft}}{\text{sec}^2}) \cdot t^2$$

$$t^2 = \frac{2 \cdot 2.9 \text{ ft}}{32 \frac{\text{ft}}{\text{sec}^2}} = 0.18$$

$$t = 0.43 \text{ sec}$$

Now we know the time it took the ball to reach the wall. At this point, we can go back to the x-dimension to finish the problem:

$$x = v_o \cdot t + \frac{1}{2} \cdot a \cdot t^2$$

$$51 \text{ ft} = (V_o) \cdot (0.43 \text{ sec}) + \frac{1}{2} \cdot (0) \cdot (0.43 \text{ sec})^2$$

$$V_o = \frac{51.0 \text{ ft}}{0.43 \text{ sec}}$$

$$V_o = 1.2 \times 10^2 \ \frac{\text{ft}}{\text{sec}}$$

This means that the ball has an initial speed of $\underline{1.2 \times 10^2 \text{ ft/sec}}$.

10. In this problem, we need to calculate the displacement in the x-dimension. As usual, however, we need a piece of information from the other dimension before we can do this. We need to use the y-dimension to figure out the time. In that dimension, the initial velocity is given by:

$$V_{o_y} = (1.2 \text{ ft / sec}) \cdot \sin(0.000) = 0$$

We also know that the acceleration is -32 m/sec^2 and that the displacement in this dimension is -4.1 ft. This is all we need to use Equation (3.19) to calculate time:

$$x = v_o \cdot t + \frac{1}{2} \cdot a \cdot t^2$$

$$-4.1 \text{ ft} = (0) \cdot t + \frac{1}{2} \cdot (-32 \ \frac{\text{ft}}{\text{sec}^2}) \cdot t^2$$

$$t^2 = \frac{2 \cdot 4.1 \text{ ft}}{32 \ \frac{\text{ft}}{\text{sec}^2}} = 0.26$$

$$t = 0.51 \text{ sec}$$

Now that we have time, we can go back and use Equation (3.19) in the x-dimension. To do this, however, we need to calculate the initial velocity in that dimension:

$$V_{o_x} = (1.2 \text{ ft / sec}) \cdot \cos(0.000) = 1.2 \text{ ft / sec}$$

Now we can use Equation (3.19):

$$x = v_o \cdot t + \frac{1}{2} \cdot a \cdot t^2$$

$$x = (1.2 \, \frac{ft}{sec}) \cdot (0.51 \, sec) + \frac{1}{2} \cdot (0) \cdot (0.51 \, sec)^2$$

$$x = 0.61 \, ft$$

So the car lands a mere 0.61 ft from the table.

MODULE 6 - ANSWERS TO THE PRACTICE PROBLEMS

1. Since there is no friction, we needn't worry about the forces that act in the vertical direction. Also, without friction, the only force involved is the one with which the father is pulling. Thus, this is a simple application of Equation (6.1):

$$\mathbf{F} = \mathbf{m} \cdot \mathbf{a}$$

$$\mathbf{F} = (25 \text{ kg}) \cdot (0.35 \frac{\text{m}}{\text{sec}^2})$$

$$\mathbf{F} = 8.8 \frac{\text{kg} \cdot \text{m}}{\text{sec}^2} = 8.8 \text{ Newtons}$$

The father, then, is pulling with a force of 8.8 Newtons.

2. Once again, without friction, this becomes a simple application of Equation (6.1). We have to be careful, however. The force is 16 pounds, but we are not given the mass. We are given the weight of the rock (501 pounds). We know that it is weight because of the unit "pounds." For Equation (6.1), we need the mass. So we will use Equation (6.2) to convert from weight to mass:

$$\text{w} = \text{m} \cdot \text{g}$$

$$501 \text{ pounds} = \text{m} \cdot (32 \frac{\text{ft}}{\text{sec}^2})$$

$$\text{m} = \frac{501 \frac{\text{slug} \cdot \text{ft}}{\text{sec}^2}}{32 \frac{\text{ft}}{\text{sec}^2}} = 16 \text{ slugs}$$

Now that we have mass, we can use Equation (6.1):

$$\mathbf{F} = \mathbf{m} \cdot \mathbf{a}$$

$$16 \text{ pounds} = (16 \text{ slugs}) \cdot \mathbf{a}$$

$$\mathbf{a} = \frac{16 \frac{\text{slug} \cdot \text{ft}}{\text{sec}^2}}{16 \text{ slugs}} = 1.0 \frac{\text{ft}}{\text{sec}^2}$$

So the rock will accelerate at 1.0 ft/sec^2.

3. If there were not friction between your feet and the ice, then the feet would slip around and could never get enough grip to allow you to push. According to Newton's Third Law, every action has an equal and opposite reaction. When you push on the rock, you use your feet to push against the ice. The ice reacts by pushing back, causing you and the rock to move. Without friction, this could never happen.

4. This is a rather simple application of Equation (6.2). We have the mass and we need to know the weight. The one thing we have to think about is the units. We really should put weight in Newtons, which contains kg. The mass is currently in grams, so we need to convert. Without writing out the conversion, the mass becomes 0.523 kg. Now we can use Equation (6.2):

$$w = m \cdot g$$

$$w = (0.523 \text{ kg}) \cdot (9.8 \frac{m}{sec^2})$$

$$w = 5.1 \frac{kg \cdot m}{sec^2} = 5.1 \text{ Newtons}$$

The weight, then, is 5.1 Newtons.

5. When an object moves from one planet to another, its weight changes but its mass doesn't. We therefore must convert from weight to mass so that we have something which is the same at both places. Since we have the weight on earth, we must use the acceleration due to gravity of earth in order to calculate the mass:

$$w = m \cdot g$$

$$1234 \text{ pounds} = m \cdot (32 \frac{ft}{sec^2})$$

$$m = \frac{1234 \frac{slug \cdot ft}{sec^2}}{32 \frac{ft}{sec^2}} = 39 \text{ slugs}$$

The mass is the same on the earth and on the moon. Thus, we can use this mass and the acceleration due to gravity on the moon to determine the weight on the moon:

$$w = m \cdot g$$

$$w = (39 \text{ slugs}) \cdot (5.3 \frac{ft}{sec^2})$$

$$w = 2.1 \times 10^2 \frac{slug \cdot ft}{sec^2} = 2.1 \times 10^2 \text{ pounds}$$

6. When an object moves from one planet to another, its weight changes but its mass doesn't. We therefore must convert from weight to mass so that we have something which is the same at both places. Since we have the weight on Mercury, we must use the acceleration due to gravity of Mercury in order to calculate the mass:

$$w = m \cdot g$$

$$296 \text{ Newtons} = m \cdot (3.95 \frac{m}{sec^2})$$

$$m = \frac{296 \frac{kg \cdot \cancel{m}}{\cancel{sec^2}}}{3.95 \frac{\cancel{m}}{\cancel{sec^2}}} = 74.9 \text{ kg}$$

The mass is the same on the earth and on Mercury. Thus, we can use this mass and the acceleration due to gravity on earth to determine the weight on the earth:

$$w = m \cdot g$$

$$w = (74.9 \text{ kg}) \cdot (9.8 \frac{m}{sec^2})$$

$$w = 7.3 \times 10^2 \frac{kg \cdot m}{sec^2} = 7.3 \times 10^2 \text{ Newtons}$$

The astronaut, then, weighs 7.3×10^2 Newtons on earth.

7. Equation (6.3) tells us that in order to calculate the frictional force, we need to know the normal force. To get that; we need the weight.

$$w = m \cdot g$$

$$w = (745 \text{ kg}) \cdot (9.8 \frac{m}{sec^2})$$

$$w = 7.3 \times 10^3 \frac{kg \cdot m}{sec^2} = 7.3 \times 10^3 \text{ Newtons}$$

This tells us that the road exerts a normal force of 7.3×10^3 Newtons. Now we can use Equation (6.3) to calculate the frictional force. We will use the coefficient of kinetic friction because the car is already moving:

$$f = \mu \cdot F_n$$

$$f = (0.32) \cdot (7.3 \times 10^3 \text{ Newtons})$$

$$f = 2.3 \times 10^3 \text{ Newtons}$$

The frictional force, then, is 2.3 x 10^3 Newtons.

8. In this problem, friction is involved. That means we'll have to sum up the forces to figure things out. To do this, we better figure out where all the forces are:

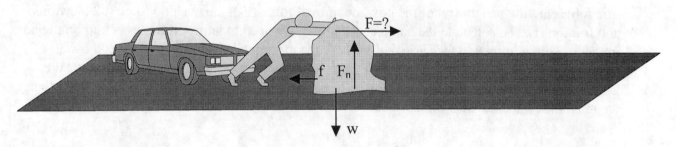

The driver is pushing the rock with a force (F=?), friction (f) opposes any motion, gravity (w) works on the rock, and the road exerts a normal force (F_n). To figure out the force that the driver pushes with, we need to determine the frictional force, which means we need the normal force. In order to get the normal force, however, we need the rock's weight. We were already given that. We were told that this is a 314 - pound rock. The unit pound means weight. Therefore, we know the weight, which means we know the normal force (314 pounds). Since we know the normal force, we can determine the frictional force using Equation (6.3). To use this equation, however, we need to determine which coefficient of friction to use. Since the driver is trying to start the rock moving, we obviously need to use the coefficient of static friction (0.45):

$$f = \mu \cdot F_n$$

$$f = (0.45) \cdot (314 \text{ pounds})$$

$$f = 1.4 \times 10^2 \text{ pounds}$$

Now that we have the frictional force, we are really done. After all, the only two forces acting in the horizontal dimension (which is the dimension in which the driver wants the motion to happen) are the frictional force and the force with which the driver is pushing. If the frictional force is 1.4 x 10^2 pounds, then the driver must apply a force greater than that. Doing so will overcome friction and get the rock moving. Thus, the driver must apply a force greater than 1.4 x 10^2 pounds.

9. We must start all problems like this one by looking at all of the forces that come into play:

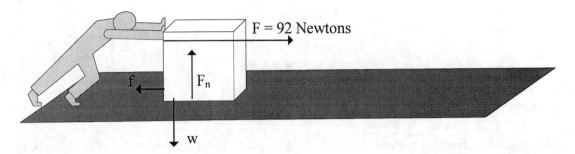

The forces are the weight (w) of the box, the normal force (F_n), the pushing force (92 Newtons), and friction (f). To calculate the box's acceleration, we need to know all forces acting on the box in the horizontal direction, because that's where the motion takes place. When we know all of those forces, we can sum them up and set them equal to the mass times the acceleration. We already know one of the two forces (92 Newtons), but we need to calculate friction. To do that, we must first calculate the normal force:

$$w = m \cdot g$$

$$w = (74 \text{ kg}) \cdot (9.8 \, \frac{m}{\sec^2})$$

$$w = 7.3 \times 10^2 \, \frac{kg \cdot m}{\sec^2} = 7.3 \times 10^2 \text{ Newtons}$$

Since the normal force counteracts the weight, we know that it also has a magnitude of 7.3×10^2 N. We can now calculate friction. To do this, we will use the coefficient of kinetic friction, because the problem tells us to make the calculation for when the box is already moving:

$$f = \mu \cdot F_n$$

$$f = (0.12) \cdot (7.3 \times 10^2 \text{ Newtons})$$

$$f = 88 \text{ Newtons}$$

Now we can *finally* sum up the horizontal forces. We need to take the directions of the force into account explicitly in our summation, so I will say that the 92 Newtons with which the man is pushing is a positive force (since it is pointed to the right) and that the frictional force is therefore negative:

$$F - f = m \cdot a$$

$$92 \text{ Newtons} - 88 \text{ Newtons} = (74 \text{ kg}) \cdot a$$

$$a = \frac{4 \frac{\text{kg} \cdot \text{m}}{\text{sec}^2}}{74 \text{ kg}} = 0.05 \frac{\text{m}}{\text{sec}^2}$$

While the man is pushing the rock, then, it accelerates at <u>0.05 m/sec²</u>.

10. As with all such problems, we need to examine all of the forces at play before we try to solve the problem:

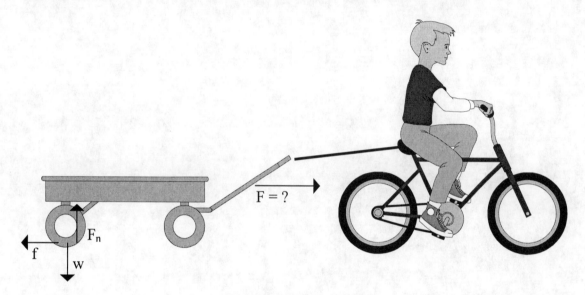

To determine the force needed, we will have to sum up the forces in the horizontal dimension. Before we do that, however, we need to determine friction. We start by determining the normal force. The wagon's weight is 11 pounds. We know that it's weight because its unit is pounds. Since the normal force counteracts weight, this means that the normal force is also 11 pounds. That's all we need to get started:

$$f = \mu \cdot F_n$$

$$f = (0.32) \cdot (11 \text{ pounds})$$

$$f = 3.5 \text{ pounds}$$

Now that we have the frictional force, we can sum up the forces in the horizontal direction and determine the force with which the bike must pull. When we sum up the forces, we set them equal to mass times acceleration. In this case, however, the acceleration is zero (constant velocity). Thus, the sum of the forces must equal zero:

$$F - f = 0$$

$$F - 3.5 \text{ pounds} = 0$$

$$F = 3.5 \text{ pounds}$$

To keep the wagon moving at a constant velocity, then, the boy must pull with a force of <u>3.5 pounds</u>.

MODULE 7 - ANSWERS TO THE PRACTICE PROBLEMS

1. If we look at the forces acting on the tightrope walker, there are three of them. Gravity is pulling down on the tightrope walker, while the tightrope pulls him up, from both sides of his foot. Thus, the force diagram looks like this:

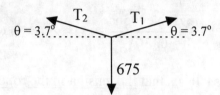

Since the tightrope walker is not falling (hopefully), he is in static equilibrium, so the sum of the forces in each dimension must be zero. So, let's split the force vectors up into their components. For T_1, the angle is 3.7 degrees, but it is not 3.7 degrees for T_2. In order to properly define the angle for vectors, you must start from the positive x-axis and move counterclockwise. Thus, the properly defined angle for T_2 is 176.3 degrees. The properly defined angle for the weight vector is 270 degrees. The vectors therefore split up in the following way:

$$T_{1x} = T_1 \cdot \cos(3.7) = 1.0 \cdot T_1$$

$$T_{1y} = T_1 \cdot \sin(3.7) = 0.065 \cdot T_1$$

$$T_{2x} = T_2 \cdot \cos(176.3) = -1.0 \cdot T_2$$

$$T_{2y} = T_2 \cdot \sin(176.3) = 0.065 \cdot T_2$$

$$w_x = (675 \text{ Newtons}) \cdot \cos(270.0) = 0$$

$$w_y = (675 \text{ Newtons}) \cdot \sin(270.0) = -675 \text{ Newtons}$$

Now that we have everything split up into x and y-components, we can sum up the forces in each dimension and make sure that they equal zero. In the x-dimension:

$$1.0 \cdot T_1 + -1.0 \cdot T_2 + 0 = 0$$

Although we can't solve for anything here, we can at least rearrange the equation:

$$T_1 = T_2$$

Now we can go to the y-dimension:

$$0.065 \cdot T_1 + 0.065 \cdot T_2 + -675 \text{ Newtons} = 0$$

This equation also has two unknowns in it, but we can use the fact that T_1 and T_2 are equal (which we learned in the y-dimension) in order to replace T_2 with T_1:

$$0.065 \cdot T_1 + 0.065 \cdot T_1 + -675 \text{ Newtons} = 0$$

$$T_1 = \frac{675 \text{ Newtons}}{0.130} = 5.19 \times 10^3 \text{ Newtons}$$

Since T_1 and T_2 are equal, this tells us that <u>the tension in the rope on each side of the tightrope walker is 5.19×10^3 Newtons</u>.

2. The force diagram in this problem is pretty simple:

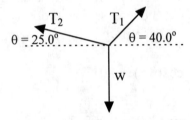

Now we need to split these vectors up into their components:

$$T_{1x} = T_1 \cdot \cos(40.0) = 0.766 \cdot T_1$$

$$T_{1y} = T_1 \cdot \sin(40.0) = 0.643 \cdot T_1$$

$$T_{2x} = T_2 \cdot \cos(155.0) = -0.9063 \cdot T_2$$

$$T_{2y} = T_2 \cdot \sin(155.0) = 0.4226 \cdot T_2$$

$$w_x = (w) \cdot \cos(270.0) = 0$$

$$w_y = (w) \cdot \sin(270.0) = -w$$

Now let's sum up these forces in each dimension and set the sum equal to zero. When we do that, I will just substitute in the weight of the sign for "w." Now remember, the problem gives us mass (15 kg), which we must convert to weight (1.5×10^2 Newtons).

$$0.766 \cdot T_1 + -0.9063 \cdot T_2 = 0$$

Although we cannot solve this equation because it has two variables, we probably realize that we will get another equation in the y dimension, and we will have to solve the equations simultaneously. Therefore, we might as well go ahead and solve for one variable in this equation in terms of the other, to get us started down that route. It doesn't matter which variable we solve for, so I choose to solve for T_2 in terms of T_1:

$$T_2 = \frac{0.766 \cdot T_1}{0.9063} = 0.845 \cdot T_1$$

Now we can go to the y dimension, sum up the forces, and set the sum equal to zero:

$$0.643 \cdot T_1 + 0.4226 \cdot T_2 + - w = 0$$

We can substitute the expression for T_2 that we got from the x-dimension, and we can also put in the weight we calculated earlier:

$$0.643 \cdot T_1 + 0.4226 \cdot (0.845 \cdot T_1) - 1.5 \times 10^2 \text{ Newtons} = 0$$

$$T_1 = 1.5 \times 10^2 \text{ Newtons}$$

We can now use that number to go back to our equation we got from the x-dimension and calculate T_2:

$$T_2 = 0.845 \cdot T_1 = 0.845 \cdot (1.5 \times 10^2 \text{ Newtons}) = 1.3 \times 10^2 \text{ Newtons}.$$

The two strings, then, have a tension of $\underline{1.5 \times 10^2 \text{ Newtons and } 1.3 \times 10^2 \text{ Newtons}}$.

3. Equation (7.1) says that the torque applied to an object is equal to the force applied times the length of the lever arm. By doubling the length of the wrench, the plumber is doubling the length of the lever arm. Since torque is calculated by multiplying force times lever arm, doubling the lever arm will double the torque. The new torque, then, will be $\underline{224 \text{ Newton·meters}}$.

4. Equation (7.1) tells us:

$$\tau = F_\perp \cdot r$$

where r is the length of the lever arm. Well, since the center of the nut is the axis of rotation, and the lever arm is defined as the distance from the axis of rotation to the force, the question is really just asking us to calculate the length of the lever arm.

$$97 \text{ Newton·meters} = 415 \text{ Newtons} \cdot r$$

$$r = \underline{0.23 \text{ meters}}$$

5. This is a rotational equilibrium problem, because in order to balance, the see-saw must be in rotational equilibrium. So, the sum of the torques must equal zero. To look at the torques, we must look at all of the forces in the problem. As was mentioned in the module, we will ignore the weight of the see-saw. If we do that, then there are only two forces to consider: the weight of the girl and the weight of the boy. If the boy were on by himself, the see-saw would tilt clockwise, so the torque that his weight applies is positive. Since we have both the lever arm distance and the weight, we can get a number for that torque:

$$\tau = (352 \text{ Newtons}) \cdot (1.12 \text{ meters}) = 394 \text{ Newton} \cdot \text{meters}$$

The only other force in the problem is the girl's weight. We cannot get a number for the torque that this applies, but we can get an equation for it. If the girl were on the see-saw by herself, it would tilt in a counter-clockwise manner, so her torque is negative:

$$\tau = - (w) \cdot (1.34 \text{ meters})$$

Now we just add the torques together and set the sum equal to zero:

$$394 \text{ Newton} \cdot \text{meters} + -(w) \cdot (1.34 \text{ meters}) = 0$$

$$w = 294 \text{ Newtons}$$

So the girl weighs 294 Newtons.

6. This problem asks for the total torque applied to the steering wheel by the captain. Well, each hand applies a torque, so all we have to do is add up the torques applied by each hand. The sum will be the total torque applied. The force applied by each hand (211 Newtons) is perpendicular to the lever arm, and the diagram illustrates the lever arm distance (1.10 meters), so to calculate the torque, we just need to use Equation (7.1):

$$\tau = F_\perp \cdot r = (211 \text{ Newtons}) \cdot (1.10 \text{ meters}) = 232 \text{ Newton} \cdot \text{meters}$$

Each hand applies this torque, because they each apply the same force at the same lever arm. The only thing we have left to do before we add these torques is determine direction. Both hands cause the wheel to move in a clockwise manner, so both torques are positive:

$$\text{Total Torque} = 232 \text{ Newton} \cdot \text{meters} + 232 \text{ Newton} \cdot \text{meters} = 464 \text{ Newton} \cdot \text{meters}$$

The captain, then, is applying 464 Newton·meters of torque to the wheel.

7. To balance the bar, all torques must sum up to zero. Since we always ignore the weight of the bar in these problems, there are only three torques. The ones caused by the 12 kg (1.2×10^2 Newton) and 21 kg (2.1×10^2 Newton) masses are negative, because they make the bar tilt counter-clockwise. The torque caused by the 18 kg (1.8×10^2 Newton) mass is positive, because

it makes the bar tilt in a clockwise manner. To calculate torque, we just take the force (which is weight- not mass) and multiply by the lever arm. The three torques are:

$$\tau_{12\ kg} = -(1.2 \times 10^2 \text{ Newtons}) \cdot (0.21 \text{ meters}) = -25 \text{ Newton·meters}$$
$$\tau_{21\ kg} = -(2.1 \times 10^2 \text{ Newtons}) \cdot (0.12 \text{ meters}) = -25 \text{ Newton·meters}$$

$$\tau_{18\ kg} = (1.8 \times 10^2 \text{ Newtons}) \cdot (x)$$

Adding these torques up and setting them equal to zero:

$$-25 \text{ Newton·meters} + -25 \text{ Newton·meters} + (1.8 \times 10^2 \text{ Newtons}) \cdot x = 0$$

$$x = 0.28 \text{ meters}$$

To balance the bar, the 18 kg mass must be placed 0.28 meters (28 cm) from the fulcrum.

8. The force diagram for this problem is rather easy, especially since there is no friction here:

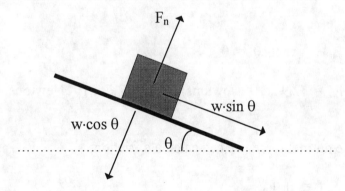

The weight of the block is 2.9×10^2 Newtons, and the angle is given as 23 degrees. Thus, the only force in the dimension that contains motion can be calculated. If we set this force equal to the mass times the acceleration, we can get our answer. We will call this force negative, since it goes down the plane:

$$-w \cdot \sin\theta = m \cdot a$$

$$-(2.9 \times 10^2 \text{ Newtons}) \cdot \sin 23 = (30.1 \text{ kg}) \cdot a$$

$$a = -3.8 \text{ m/sec}^2$$

So the acceleration is down the ramp (negative was defined as down the ramp) at 3.8 m/sec².

9. Reconsidering the above problem with friction, the force diagram looks like this:

Now we have two forces to consider. One of them is friction, so we might as well go to the perpendicular dimension and determine what the friction is. In that dimension, the sum of the forces needs to equal zero, because there is no movement in that dimension:

$$F_n + -w \cdot \cos\theta = 0$$

$$F_n = (2.9 \times 10^2 \text{ Newtons}) \cdot \cos 23 = 2.7 \times 10^2 \text{ Newtons}$$

Since we now have the normal force, we can calculate friction:

$$f = \mu_k \cdot F_n$$

$$f = (0.35) \cdot (2.7 \times 10^2 \text{ Newtons}) = 95 \text{ Newtons}$$

Now that we have the frictional force, we can finally look at the dimension that contains the motion:

$$f + -w \cdot \sin\theta = m \cdot \mathbf{a}$$

$$95 \text{ Newtons} + -(2.9 \times 10^2 \text{ Newtons}) \cdot \sin 23 = (30.1 \text{ kg}) \cdot \mathbf{a}$$

$$\mathbf{a} = -0.61 \text{ m/sec}^2$$

When friction is taken into account, then, the acceleration drops to <u>0.61 m/sec^2 down the ramp</u>.

10. When dealing with multiple objects, you have to consider them individually. The force diagram for the rear ship looks like this:

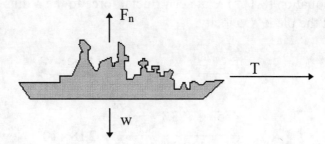

because the tension in the string is the only horizontal force acting on the ship at the rear. Thus, this one force equals the ships mass times its acceleration:

$$T = (0.45 \text{ kg}) \cdot a$$

There are two unknowns in this equation, so we need to look at the other ship for a second equation. This ship has two horizontal forces acting on it: the tension in the string and the child's pull:

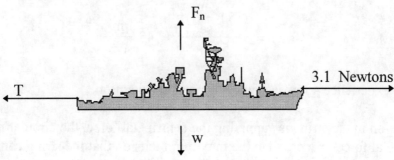

These forces must also sum to the mass of this ship times the acceleration:

$$-T + 3.1 \text{ Newtons} = (0.32 \text{ kg}) \cdot a$$

This equation also has two unknowns, but we can substitute the expression for T that we found using the other ship:

$$-(0.45 \text{ kg}) \cdot a + 3.1 \text{ Newtons} = (0.32 \text{ kg}) \cdot a$$

$$a = 4.0 \text{ m/sec}^2$$

Now this is not the only answer. The question also asked for the tension in the string. So we have to take this acceleration and put it back into the equation we got from the first ship. That will give us the tension:

$$T = (0.45 \text{ kg}) \cdot a = (0.45 \text{ kg}) \cdot (4.0 \text{ m/sec}^2) = 1.8 \text{ Newtons}$$

So the string has a tension of <u>1.8 Newtons</u> and the ships accelerate at <u>4.0 m/sec^2</u>.

MODULE 8 - ANSWERS TO THE PRACTICE PROBLEMS

1. We can use Equation (8.1) to relate centripetal force to the radius of the circular path that the plane travels and the plane's velocity:

$$F_c = \frac{m \cdot v^2}{r}$$

$$F_c = \frac{455\,\text{kg} \cdot (73.3\,\frac{m}{\sec})^2}{112.2\,m} = 2.18 \times 10^4 \text{ Newtons}$$

So the plane experiences the incredible force of 2.18×10^4 Newtons when executing this stunt.

2. This problem is a straightforward application of Equation (8.2):

$$a_c = \frac{v^2}{r}$$

$$a_c = \frac{(15.1\,\frac{m}{\sec})^2}{5.1\,m} = 45\,\frac{m}{\sec^2}$$

The riders, then, experience 45 m/sec^2 in centripetal acceleration.

3. Since the tension in the string is supplying the centripetal force, the circular motion cannot require a centripetal force of more than the maximum tension that the string can withstand. The faster the airplane twirls, the more centripetal force it will require to continue moving in a circle. Thus, we need to say that the speed of the object at the moment it broke the string determines the tension at which the string breaks. We can use Equation (8.1) to determine what that tension is.

Before we do that, however, we need to look at the units. The radius is given in cm and the mass is given in grams. The tension, however, is given in Newtons. The unit Newton uses kg and m. Thus, at this point, our units are not consistent. I will fix that by converting cm into m and g into kg. I assume that you know how to do that at this point.

$$F_c = \frac{m \cdot v^2}{r}$$

$$F_c = \frac{(0.0150\,\text{kg}) \cdot (12.1\,\frac{m}{\sec})^2}{0.345\,m}$$

$$F_c = 6.37 \text{ Newtons}$$

That's the tension that caused the string to break. Thus, the string can withstand any tension less than 6.37 Newtons.

4. In order to solve problem, we must first develop an expression for the friction that exists between the tires and the road. The force of friction depends on the normal force (which for flat surfaces is just the weight of the car) times the coefficient of friction. We do not know the mass or weight of the car, so we cannot get a number for the frictional force. We can, however, develop an expression:

$$f = \mu_k \cdot F_n$$

$$f = (0.22) \cdot (m) \cdot (9.8 \frac{\text{meters}}{\text{sec}^2})$$

$$f = 2.2 \frac{\text{meters}}{\text{sec}^2} \cdot (m)$$

Now, even though we don't have a number for the frictional force, let's go ahead and put the expression we have for it into Equation (8.1). The expression we have for the frictional force represents the maximum possible centripetal force available. If we put that into Equation (8.1), we can use it to determine the maximum speed that the car can have:

$$F_c = \frac{m \cdot v^2}{r}$$

$$2.2 \frac{\text{meters}}{\text{sec}^2} \cdot (m) = \frac{m \cdot v^2}{13.4 \text{ meters}}$$

Since mass appears on both sides of the equation, it cancels out. Thus, it doesn't matter that we don't know the mass of the car, because it cancels out of the final equation. Now we can simply solve for v:

$$2.2 \frac{\text{meters}}{\text{sec}^2} \cdot (\cancel{m}) = \frac{\cancel{m} \cdot v^2}{13.4 \text{ meters}}$$

$$v = 5.4 \frac{\text{meters}}{\text{sec}}$$

So the car can only travel at a speed of 5.4 m/sec. Since this is only about 11 mph, the driver had better be careful!

5. Since frequency is defined as the number of times per second that an object completes a revolution, the child is essentially measuring frequency. However, we need to get it into the right

units. Frequency is measure in 1/sec, but the child counts the number of revolutions per minute. Thus, the frequency is:

$$f = \frac{10}{60 \text{ sec}} = 0.17 \frac{1}{\text{sec}}$$

So the <u>train's frequency is 0.17 Hz</u>. To get the period, we simply use Equation (8.4):

$$f = \frac{1}{T}$$

$$T = \frac{1}{f} = \frac{1}{0.17 \frac{1}{\text{sec}}} = 5.9 \text{ sec}$$

<u>The period, then, is 5.9 sec</u>. If you had worked out the period first and then used it to find frequency, you would have gotten 6.0 seconds for the period. The difference between that answer and this one is due to rounding errors.

6. This is a straightforward application of Equation (8.3), as long as we keep our units straight. The centimeters will need to be converted to meters in order to keep the units for r consistent with the units in G.

$$F_g = \frac{G \cdot m_1 \cdot m_2}{r^2}$$

$$F_g = \frac{(6.67 \times 10^{-11} \frac{\text{Newton} \cdot \text{m}^2}{\text{kg}^2}) \cdot (15.0 \text{ kg}) \cdot (25.0 \text{ kg})}{(0.45 \text{ m})^2} = 1.2 \times 10^{-7} \text{ Newtons}$$

7. To figure out how long it takes Saturn to make on orbit around the sun (its orbital period, in other words), we need to determine how fast it travels. To do this, we must set the gravitational force equal to the centripetal force. Since Saturn is traveling in a circle, the mass term in Equation (8.1) refers to the mass of Saturn.

$$F_g = F_c$$

$$\frac{G \cdot m_{sun} \cdot m_{Saturn}}{r^2} = \frac{m_{Saturn} \cdot v^2}{r}$$

$$\frac{(6.67 \times 10^{-11} \frac{\text{Newton} \cdot \text{m}^2}{\text{kg}^2}) \cdot (2.0 \times 10^{30} \text{ kg})}{1.4 \times 10^{12} \text{ m}} = v^2$$

$$v = 9.8 \times 10^3 \frac{\text{m}}{\text{sec}}$$

This, of course, is not our answer. To calculate period, we need to determine how far Saturn travels in one orbit. Given the orbital radius, this is no problem:

$$\text{circumference} = 2 \cdot \pi \cdot (1.4 \times 10^{12} \text{ m}) = 8.8 \times 10^{12} \text{ m}$$

That's the distance traveled in one orbit. Now we can calculate the period:

$$\text{distance} = \text{rate} \cdot \text{time}$$

$$\text{time} = \frac{8.8 \times 10^{12} \text{ m}}{9.8 \times 10^3 \frac{\text{m}}{\text{sec}}} = 9.0 \times 10^8 \text{ sec}$$

This tells us that it takes Saturn 9.0×10^8 sec (29 years) to make one orbit around the sun.

8. We know that the orbital period of the rock is 3.9×10^6 seconds. Since we know that our equations do not use period, we might as well convert this to speed right away:

$$\text{circumference} = 2 \cdot \pi \cdot r = 2 \cdot \pi \cdot (5.0 \times 10^8 \text{ m}) = 3.1 \times 10^9 \text{ m}$$

$$\text{speed} = \frac{3.1 \times 10^9 \text{ m}}{3.9 \times 10^6 \text{ sec}} = 7.9 \times 10^2 \frac{\text{m}}{\text{sec}}$$

Now that we have speed, we can set the gravitational force equal to the centripetal force and solve for the mass of Venus:

$$F_g = F_c$$

$$\frac{G \cdot m_{\text{Venus}} \cdot m_{\text{rock}}}{r^2} = \frac{m_{\text{rock}} \cdot v^2}{r}$$

$$\frac{(6.67 \times 10^{-11} \frac{\text{Newton} \cdot \text{m}^2}{\text{kg}^2}) \cdot (m_{Venus})}{5.0 \times 10^8 \, \text{m}} = (7.9 \times 10^2 \, \frac{\text{m}}{\text{sec}})^2$$

$$m_{Venus} = 4.7 \times 10^{24} \, \text{kg}$$

The mass of the Venus is $\underline{4.7 \times 10^{24} \, \text{kg}}$.

9. There is only one orbital radius that will allow any given satellite speed. To determine the radius, we simply set the gravitational force equal to the centripetal force, as we always have:

$$F_g = F_c$$

$$\frac{G \cdot m_{earth} \cdot m_{satellite}}{r^2} = \frac{m_{satellite} \cdot v^2}{r}$$

$$\frac{(6.67 \times 10^{-11} \frac{\text{Newton} \cdot \text{m}^2}{\text{kg}^2}) \cdot (5.98 \times 10^{24} \, \text{kg})}{(1123 \, \frac{\text{m}}{\text{sec}})^2} = r$$

$$r = 3.16 \times 10^8 \, \text{m}$$

The satellite orbits the earth with a radius of $\underline{3.16 \times 10^8 \, \text{m}}$.

10. To get the orbital period, we will first need to determine how fast the satellite travels. To do that, we will need its orbital radius. The problem does not give us that. It gives us the altitude. Therefore, the first thing that we must do is calculate the orbital radius. Note that the earth's radius and the altitude are in different units. I will convert km to m to correct for this.

orbital radius = radius of earth + altitude

orbital radius = 6.38×10^6 m + 2.314×10^6 m = 8.69×10^6 m

Now we can set the gravitational force equal to the centripetal force and solve for speed:

$$F_g = F_c$$

$$\frac{G \cdot m_{earth} \cdot \cancel{m_{satellite}}}{r^2} = \frac{\cancel{m_{satellite}} \cdot v^2}{\cancel{r}}$$

$$\frac{(6.67 \times 10^{-11} \frac{Newton \cdot m^2}{kg^2}) \cdot (5.98 \times 10^{24} \cancel{kg})}{(8.69 \times 10^6 \cancel{m})} = v^2$$

$$v = 6.77 \times 10^3 \frac{m}{sec}$$

Now this is not the answer. The question asked for orbital period of the satellite. To get that, we will need to calculate how far the satellite travels in one orbit ($2\pi r$) and divide it by the speed we just determined:

$$\text{circumference} = 2 \cdot \pi \cdot (8.69 \times 10^6 \text{ m}) = 5.46 \times 10^7 \text{ m}$$

$$\text{period} = \frac{\text{distance traveled}}{\text{speed}} = \frac{5.46 \times 10^7 \cancel{m}}{6.77 \times 10^3 \frac{\cancel{m}}{sec}} = 8.06 \times 10^3 \text{ sec}$$

The period of the satellite, then, is <u>8.06×10^3 sec (2.2 hours)</u>.

MODULE 9 - ANSWERS TO THE PRACTICE PROBLEMS

1. In this problem, we are given the force that friction uses and the distance over which the force is applied. Thus, this is a simple application of Equation (9.1):

$$W = F \cdot x$$

$$W = (12.2 \text{ Newtons}) \cdot (11.5 \text{ m}) = 1.40 \times 10^2 \text{ J}$$

The worker does $\underline{1.40 \times 10^2 \text{ J}}$ of work.

2. Calculating the potential energy is a simple application of Equation (9.2):

$$PE = m \cdot g \cdot h$$

$$PE = (567 \text{ kg}) \cdot (9.8 \frac{m}{\sec^2}) \cdot (15.1 \text{ m}) = 8.4 \times 10^4 \text{ J}$$

Calculating the kinetic energy is also a simple application of Equation (9.3):

$$KE = \frac{1}{2} \cdot m \cdot v^2$$

$$KE = \frac{1}{2} \cdot (567 \text{ kg}) \cdot (19.1 \frac{m}{\sec})^2$$

$$KE = 1.03 \times 10^5 \text{ J}$$

Equation (9.4) tells us that the total energy is the sum of the two:

$$TE = PE + KE = 8.4 \times 10^4 \text{ J} + 1.03 \times 10^5 \text{ J} = 1.87 \times 10^5 \text{ J}$$

Therefore, the car's potential energy is 8.4×10^4 J, its kinetic energy is 1.03×10^5 J and its total energy is 1.87×10^5 J.

3. When the roller coaster car sits at the top of the first hill, it has no kinetic energy, because it is not moving. Thus, all of its energy is potential. Since we know the height of the hill, we can calculate that potential energy:

$$PE = m \cdot g \cdot h$$

$$PE = (124 \text{ kg}) \cdot (9.8 \frac{m}{\sec^2}) \cdot (75 \text{ m}) = 9.1 \times 10^4 \text{ J}$$

Now remember, this is also the *total* energy that the car has, because it has no kinetic energy at the top of the hill. As a result, when the car hits the top of the next hill, the total energy is still 9.1 x 10^4 J. Since the car is still moving, at this point, it also has kinetic energy. All we know is that the sum of the potential energy and kinetic energy must be 9.1 x 10^4 J. Since we have the height of the hill, however, we can calculate the potential energy:

$$PE = m \cdot g \cdot h$$

$$PE = (124 \text{ kg}) \cdot (9.8 \frac{m}{\sec^2}) \cdot (40.0 \text{ m}) = 4.9 \times 10^4 \text{ J}$$

Now that we have the potential energy, we can use Equation (9.4) to calculate the kinetic energy:

$$TE = PE + KE$$

$$9.1 \times 10^4 \text{ J} = 4.0 \times 10^4 \text{ J} + KE$$

$$KE = 9.1 \times 10^4 \text{ J} - 4.9 \times 10^4 \text{ J} = 4.2 \times 10^4 \text{ J}$$

Now that we have the kinetic energy, we can finally calculate the speed:

$$KE = \frac{1}{2} \cdot m \cdot v^2$$

$$4.2 \times 10^4 \text{ J} = \frac{1}{2} \cdot (124 \text{ kg}) \cdot v^2$$

$$v = \sqrt{\frac{2 \cdot 4.2 \times 10^4 \text{ J}}{124 \text{ kg}}} = 26 \frac{m}{\sec}$$

At the top of the second hill, the roller coaster car is traveling at 26 m/sec.

4. This problem is much like the roller coaster problem above, but the car actually has some kinetic energy at the top of the hill. Thus, we need to add potential energy and kinetic energy in order to get the total energy of the car. Since we know the height of the hill, we can calculate the potential energy:

$$PE = m \cdot g \cdot h$$

$$PE = (341 \text{ kg}) \cdot (9.8 \frac{m}{\sec^2}) \cdot (11.5 \text{ m}) = 3.8 \times 10^4 \text{ J}$$

Since we have the car's speed, we can also calculate its kinetic energy:

$$KE = \frac{1}{2} \cdot m \cdot v^2$$

$$KE = \frac{1}{2} \cdot (341 \text{ kg}) \cdot (21.4 \frac{m}{\sec})^2$$

$$KE = 7.8 \times 10^4 \text{ J}$$

The car's total energy, then, is the sum of the two:

$$TE = KE + PE = 3.8 \times 10^4 \text{ J} + 7.8 \times 10^4 \text{ J} = 1.16 \times 10^5 \text{ J}$$

Since the car coasts, the total energy does not change. As a result, when the car hits the top of the next hill, the total energy is still 1.16×10^5 J. We know, then, that the sum of the potential energy and kinetic energy must be 1.16×10^5 J. Since we have the height of the hill, we can calculate the potential energy:

$$PE = m \cdot g \cdot h$$

$$PE = (341 \text{ kg}) \cdot (9.8 \frac{m}{\sec^2}) \cdot (5.1 \text{ m}) = 1.7 \times 10^4 \text{ J}$$

Now that we have the potential energy, we can use Equation (9.4) to calculate the kinetic energy:

$$TE = PE + KE$$

$$1.16 \times 10^5 \text{ J} = 1.7 \times 10^4 \text{ J} + KE$$

$$KE = 1.16 \times 10^5 \text{ J} - 1.7 \times 10^4 \text{ J} = 9.9 \times 10^4 \text{ J}$$

Now that we have the kinetic energy, we can finally calculate the speed:

$$KE = \frac{1}{2} \cdot m \cdot v^2$$

$$9.9 \times 10^4 \text{ J} = \frac{1}{2} \cdot (341 \text{ kg}) \cdot v^2$$

$$v = \sqrt{\frac{2 \cdot 9.9 \times 10^4 \text{ J}}{341 \text{ kg}}} = 24 \frac{\text{m}}{\text{sec}}$$

At the top of the second hill, the car is traveling at 24 m/sec.

5. When the marksman fires the rifle, the bullet starts speeding through the air. Thus, it has kinetic energy. When it hits the pendulum and sticks, that kinetic energy is transferred to the pendulum. With this new energy, the pendulum starts to move up. When it reaches its maximum height (1.8 m), it has converted all of its kinetic energy into potential energy. Thus, when we calculate the potential energy of the pendulum at its maximum height, we will know the initial kinetic energy that the pendulum had. This is the same as the initial energy that the bullet had. From that we can calculate speed. Let's start, then, by calculating the potential energy of the pendulum at 1.8 m. Since we are calculating the potential energy of the pendulum, we must use the pendulum's mass (15 kg):

$$PE = m \cdot g \cdot h$$

$$PE = (15 \text{ kg}) \cdot (9.8 \frac{\text{m}}{\text{sec}^2}) \cdot (1.8 \text{ m}) = 2.6 \times 10^2 \text{ J}$$

As we just discussed, this must be equal to the kinetic energy of the pendulum, which is equal to the initial kinetic energy of the bullet. So, we can use this energy to calculate the speed of the bullet. Since we are interested in the bullet this time, however, we must use the bullet's mass (0.121 kg):

$$KE = \frac{1}{2} \cdot m \cdot v^2$$

$$2.6 \times 10^2 \text{ J} = \frac{1}{2} \cdot (0.121 \text{ kg}) \cdot v^2$$

$$v = \sqrt{\frac{2 \cdot 2.6 \times 10^2 \text{ J}}{0.121 \text{ kg}}} = 66 \frac{\text{m}}{\text{sec}}$$

The bullet had an initial speed of 66 m/sec.

6. Since the toy comes to a halt, all energy has been removed by friction. The energy it had when it started is equal to the kinetic energy of the toy:

$$KE = \frac{1}{2} \cdot m \cdot v^2$$

$$KE = \frac{1}{2} \cdot (0.124 \text{ kg}) \cdot (3.1 \frac{m}{\sec})^2$$

$$KE = 0.60 \text{ J}$$

Since that's the original amount of energy that the toy had, it must all be removed by friction in order to get the toy to stop. The only way that can happen is if friction does 0.60 J of work.

7. Since the package is sitting still at the top of the ramp, its total energy is the same as its potential energy. Since we have the height and the mass, we can calculate it:

$$PE = m \cdot g \cdot h$$

$$PE = (1.1 \text{ kg}) \cdot (9.8 \frac{m}{\sec^2}) \cdot (1.2 \text{ m}) = 13 \text{ J}$$

If friction were not involved, how much total energy would the package have at the bottom of the hill? It would have 13 J, all of which would be kinetic. Since friction is working on the package, however, its total energy will be less. How much less? Well, we have the speed and mass of the package, so we can calculate it:

$$KE = \frac{1}{2} \cdot m \cdot v^2$$

$$KE = \frac{1}{2} \cdot (1.1 \text{ kg}) \cdot (2.1 \frac{m}{\sec})^2$$

$$KE = 2.4 \text{ J}$$

Now, since the box has no height, this energy is also its total energy. Thus, during the fall, the box went from a total energy of 13 J to a total energy of 2.4 J. There were 11 J lost. Where did those 11 J go? Friction converted them into heat. The only way friction could do that was by working, so friction did 11 J of work.

8. When the student gives the book a shove, he gives it kinetic energy:

$$KE = \frac{1}{2} \cdot m \cdot v^2$$

$$KE = \frac{1}{2} \cdot (0.534 \text{ kg}) \cdot (4.1 \frac{m}{\sec})^2$$

$$KE = 4.5 \text{ J}$$

Since the table is level, we can ignore potential energy. Thus, the total energy of the book is 4.5 J. In order to come to a halt, the book must lose all of that energy to friction. This means that friction must do 4.5 J of work. Since we know the coefficient of kinetic friction, we can calculate the force with which friction works:

$$f = \mu \cdot F_n$$

$$f = (0.45) \cdot (0.534 \text{ kg}) \cdot (9.8 \frac{m}{\sec^2}) = 2.4 \text{ Newtons}$$

Now that we know the work done and the force, we can calculate the distance:

$$W = F \cdot x$$

$$4.5 \text{ J} = (2.4 \text{ Newtons}) \cdot x$$

$$x = \frac{4.5 \cancel{\text{Newton}} \cdot m}{2.4 \cancel{\text{Newtons}}} = 1.9 \text{ m}$$

The book, therefore, slides <u>1.9 m</u> before coming to a halt.

9. Power is work divided by time. Since we have force and distance, we can calculate work:

$$W = F \cdot x$$

$$W = (67 \text{ Newtons}) \cdot (11.1 \text{ m}) = 7.4 \times 10^2 \text{ J}$$

Once we convert time into seconds, we can plug this work into Equation (9.5):

$$P = \frac{W}{t}$$

$$P = \frac{7.4 \times 10^2 \text{ J}}{144 \text{ sec}} = \underline{5.1 \text{ Watts}}$$

10. We are given the Wattage of the bulb and the time it burns. Using Equation (9.5), we can therefore calculate the work that could be done:

$$P = \frac{W}{t}$$

$$101 \text{ Watts} = \frac{W}{3600 \text{ sec}}$$

$$W = (101 \frac{\text{J}}{\text{sec}}) \cdot (3600 \text{ sec}) = 3.6 \times 10^5 \text{ J}$$

Now that we know how much work can be generated, Equation (9.1) will tell us the force:

$$W = F \cdot x$$

$$3.6 \times 10^5 \text{ J} = F \cdot (25 \text{ m})$$

$$F = \frac{3.6 \times 10^5 \text{ Newton} \cdot \text{m}}{25 \text{ m}} = 1.4 \times 10^4 \text{ Newtons}$$

That amount of power could generate a force of $\underline{1.4 \times 10^4 \text{ Newtons}}$ over a distance of 25 meters.

MODULE 10 - ANSWERS TO THE PRACTICE PROBLEMS

1. We know the mass and momentum, and we need to calculate the velocity. Thus, this problem uses Equation (10.1). The negative sign in momentum depicts direction, so we had better keep it in the equation to keep track of direction.

$$\mathbf{p} = m \cdot \mathbf{v}$$

$$-9.6 \frac{kg \cdot m}{sec} = (4.6 \text{ kg}) \cdot \mathbf{v}$$

$$\mathbf{v} = -2.1 \frac{m}{sec}$$

Notice that weight (Newtons) had to be converted to mass (kg) in order to determine that the child has a velocity of <u>-2.1 m/sec</u>. Once again, the negative sign merely denotes direction.

2. This is a simple application of Equation (10.1). It is put here to emphasize that direction is an essential part of momentum. The direction of the velocity vector is given. Even though we cannot put it in the equation, we must remember it so that we can attach it to the momentum when we report the answer:

$$\mathbf{p} = m \cdot \mathbf{v}$$

$$\mathbf{p}_{final} = (0.0251 \text{ kg}) \cdot (351 \frac{m}{sec}) = 8.81 \frac{kg \cdot m}{sec}$$

Notice we had to convert grams to kg to determine that <u>the bullet's momentum is 8.81 $\frac{kg \cdot m}{sec}$ at 45 degrees northeast</u>. Without the direction, the answer would be wrong!

3. In this problem, the golf ball has zero momentum before the club hits it. To calculate the change in momentum, we must calculate the momentum after the hit. Notice that we need to convert grams to kg in order to calculate the momentum.

$$\mathbf{p} = m \cdot \mathbf{v}$$

$$\mathbf{p} = (0.046 \text{ kg}) \cdot (55 \frac{m}{sec}) = 2.5 \frac{kg \cdot m}{sec}$$

Since the initial momentum was zero, the change in momentum is the same as the final momentum. Now that we know the change in momentum, we can combine that with the time interval (0.090 seconds) and Equation (10.6) to give us the force used:

$$F = \frac{\Delta p}{\Delta t}$$

$$F = \frac{2.5 \frac{kg \cdot m}{sec}}{0.090 \, sec} = 28 \frac{kg \cdot m}{sec^2} = 28 \, \text{Newtons}$$

The club, then, exerted a force of <u>28 Newtons</u>.

4. In order to stop the clay, the tree must exert an impulse. We can determine it by figuring out the change in momentum. We know that the final momentum is zero, but we need to calculate the initial momentum.

$$p = m \cdot v$$

$$p = (0.456 \, kg) \cdot (5.9 \frac{m}{sec}) = 2.7 \frac{kg \cdot m}{sec}$$

Notice that since only the speed was given, we cannot use vector notation. Only the magnitudes of the vectors are being considered here. Notice also that the mass had to be converted from grams to kg in order for us to calculate the momentum.

Now we know the change in momentum. If the clay was traveling with a momentum of 2.7 $\frac{kg \cdot m}{sec}$ and its momentum changed to zero, the change in momentum (final minus initial) is -2.7 $\frac{kg \cdot m}{sec}$. We will have to drop the negative sign, however, because we cannot deal with vectors in the problem. Now we know the change in momentum as well as the force applied, so we can calculate the time interval.

$$F = \frac{\Delta p}{\Delta t}$$

$$4.0 \, \text{Newtons} = \frac{2.7 \frac{kg \cdot m}{sec}}{\Delta t}$$

$$\Delta t = \frac{2.7 \frac{kg \cdot m}{sec}}{4.0 \frac{kg \cdot m}{sec^2}} = 0.68 \ sec$$

The tree, then, takes <u>0.68 seconds</u> to stop the clay.

5. In order to determine the recoil velocity, we must calculate the total momentum of the system both before and after the gun was fired. Since the Law of Momentum Conservation says that they must equal each other, we can build the following equation:

$$(m_{gun} \cdot v_{gun} + m_{shell} \cdot v_{shell})_{before} = (m_{gun} \cdot v_{gun} + m_{shell} \cdot v_{shell})_{after}$$

Since both the gun and the shell are stationary, both momenta are equal to zero before the shot. We have all of the rest of the information in the equation except for the velocity of the gun after the shot was fired. We can therefore solve for it.

$$0 + 0 = (975 \ kg) \cdot v_{gun} + (85 \ kg) \cdot (345 \frac{m}{sec})$$

$$0 = (975 \ kg) \cdot v_{gun} + 2.9 \times 10^4 \frac{kg \cdot m}{sec}$$

$$v_{gun} = \frac{-2.9 \times 10^4 \frac{kg \cdot m}{sec}}{975 \ kg} = -3.0 \times 10^1 \frac{m}{sec}$$

The negative sign simply indicates that the gun travels in the opposite direction as compared to the shell. Thus, its recoil velocity is <u>3.0×10^1 m/sec away from the shell</u>.

6. Since the sum of the forces acting on our system (the skater and the bowling ball) is zero, we can simply calculate the momenta before and after the throw and set them equal to each other.

$$(m_{bowlingball} \cdot v_{bowlingball} + m_{skater} \cdot v_{skater})_{before} = (m_{bowlingball} \cdot v_{bowlingball} + m_{skater} \cdot v_{skater})_{after}$$

The skater and ball stand still initially, so their initial velocities (and therefore their initial momenta) are zero.

$$0 + 0 = (6.0 \ kg) \cdot (-3.0 \frac{m}{sec}) + (78 \ kg) \cdot v_{skater}$$

$$0 = -18 \frac{kg \cdot m}{sec} + (78 \ kg) \cdot v_{skater}$$

$$v_{skater} = \frac{18 \frac{kg \cdot m}{sec}}{78 \text{ kg}} = 0.23 \frac{m}{sec}$$

The lack of a negative sign simply means that the skater travels in the <u>opposite direction as the ball at 0.23 m/sec</u>.

7. In this problem, we have two independently moving railroad cars at first. When they collide, however, they will stick together. This means that they now can be treated as one single object. Since we can ignore friction, and since the force due to gravity working on the cars is canceled by the normal force of the rail pushing against them, we can say that momentum is conserved. Thus, we can say that the total momentum before the collision must be the same as the total momentum afterwards

$$(m_1 \cdot v_1 + m_2 \cdot v_2)_{before} = (m_1 + m_2) \cdot v_{both}$$

We know everything in this equation except the velocity of the cars when they travel together, so we can solve for it:

$$(0.20 \text{ kg}) \cdot (0.24 \frac{m}{sec}) + (0.42 \text{ kg}) \cdot (0.45 \frac{m}{sec}) = (0.20 \text{ kg} + 0.42 \text{ kg}) \cdot v_{both}$$

$$0.24 \frac{kg \cdot m}{sec} = (0.62 \text{ kg}) \cdot v_{both}$$

$$v_{both} = \frac{0.24 \frac{kg \cdot m}{sec}}{0.62 \text{ kg}} = 0.39 \frac{m}{sec}$$

Since the velocity is positive, that means that the two cars <u>travel with a velocity of 0.39 m/sec in their initial directions</u>.

8. In this problem, the object under consideration (the freight car) changes its mass. Since mass is a part of momentum, this affects the car's momentum. Assuming momentum is conserved, we can develop the following equation:

$$m_{empty} \cdot v_{empty} = m_{full} \cdot v_{full}$$

We are told the mass and velocity of the car when it is empty. In addition, we are also told the mass of the junked car that is put inside. Thus, we can say that this mass of junked car plus the mass of the empty freight car is the mass of the freight car when it is full. The only thing that we don't know is the velocity of the freight car when it is full, so we can solve for it.

$$(895 \text{ kg}) \cdot (3.1 \frac{m}{sec}) = (895 \text{ kg} + 365 \text{ kg}) \cdot v_{full}$$

$$v_{full} = \frac{2.8 \times 10^3 \frac{kg \cdot m}{sec}}{1260 \text{ kg}} = 2.2 \frac{m}{sec}$$

The freight car, because of the added mass, must slow to a velocity of 2.2 m/sec in order to conserve momentum.

9. Angular momentum is given by Equation (10.11):

$$L = m \cdot v \cdot r$$

$$4.5 \frac{kg \cdot m^2}{sec^2} = (0.25 \text{ kg}) \cdot (3.5 \frac{m}{sec}) \cdot r$$

$$r = 5.1 \text{ m}$$

The radius of motion, then is 5.1 meters.

10. Once the rock starts twirling, it has angular momentum. When the string is adjusted, the radius of the motion changes, but the angular momentum cannot, due to the Law of Angular Momentum Conservation. As a result, the angular momentum before the radius changes must be the same as the angular momentum after it changes:

$$(m \cdot v \cdot r)_{before} = (m \cdot v \cdot r)_{after}$$

Since the mass of the rock does not change, it is the same on both sides of the equation, so it cancels out. We know everything else but the radius afterwards, so we can solve for it:

$$\cancel{m} \cdot (5.6 \frac{meters}{sec}) \cdot (0.34 \text{ meters}) = \cancel{m} \cdot (3.2 \frac{m}{sec}) \cdot r_{after}$$

$$r_{after} = \frac{(5.6 \frac{\cancel{meters}}{\cancel{sec}}) \cdot (0.34 \text{ meters})}{3.2 \frac{\cancel{meters}}{\cancel{sec}}} = 0.60 \text{ meters}$$

The ball's new radius of motion is 0.60 meters, or 60 cm.

MODULE 11 - ANSWERS TO THE PRACTICE PROBLEMS

1. In this problem, we are given the weight of the bag of vegetables, which is the force due to gravity. Since gravity pulls down, we will say that this is a negative force. Thus, gravity pulls the spring with a force of -10.0 Newtons. In order to keep the bag from falling, the spring exerts a restoring force equal to but opposite of the force due to gravity. Thus, the spring exerts a force of +10.0 Newtons. This is **F** in Equation (11.1).

When the bag hangs on the spring, it stretches downwards by 5.6 cm. Thus, the displacement (**x**) is -0.056 m, after converting to standard units. Using the force calculated above and this displacement, we can use Equation (11.1) to determine the force constant of the spring:

$$\mathbf{F} = -\mathbf{k} \cdot \mathbf{x}$$

$$10.0 \text{ Newtons} = -\mathbf{k} \cdot (-0.056 \text{ m})$$

$$\mathbf{k} = \frac{10.0 \text{ Newtons}}{0.056 \text{ m}} = 1.8 \times 10^2 \frac{\text{Newtons}}{\text{m}}$$

The spring constant, therefore, is <u>1.8 x 10² Newtons/meter</u>.

2. This is another problem using Equation (11.1), but it is a little more complicated than the first one. This time, we must calculate acceleration. How do we do that? Well, first, we use Equation (11.1) to get the force that the dart experiences. Then, since we have the mass of the dart, we can use our old friend **F**=m·**a** to determine the acceleration. We start, then, by calculating the force experienced by the dart:

$$\mathbf{F} = -\mathbf{k} \cdot \mathbf{x}$$

$$\mathbf{F} = -52 \, \frac{\text{Newtons}}{\cancel{\text{m}}} \cdot (0.043 \, \cancel{\text{m}})$$

$$\mathbf{F} = -2.2 \text{ Newtons}$$

What does the negative sign mean? It denotes direction. Since we assumed that the displacement was positive, the negative sign just means that the force which the dart experiences is opposite the direction of the displacement. Now that we have force, we can get acceleration:

$$\mathbf{F} = \mathbf{m} \cdot \mathbf{a}$$

$$-2.2 \text{ Newtons} = (0.075 \text{ kg}) \cdot \mathbf{a}$$

$$a = \frac{-2.2 \text{ Newtons}}{0.075 \text{ kg}} = \frac{-2.2 \frac{\text{kg} \cdot \text{m}}{\text{sec}^2}}{0.075 \text{ kg}} = -29 \frac{\text{m}}{\text{sec}^2}$$

The negative sign simply means that the dart is experiencing acceleration opposite that of the spring's compression. The acceleration, then, is -29 m/sec².

3. In this problem, we are given the mass and the spring constant, and we are asked to calculate the period. This is a direct application of Equation (11.11):

$$T = 2 \cdot \pi \sqrt{\frac{m}{k}}$$

$$T = 2 \cdot \pi \sqrt{\frac{0.050 \text{ kg}}{1.5 \frac{\text{Newtons}}{\text{m}}}} = 2 \cdot \pi \sqrt{\frac{0.050 \text{ kg}}{1.5 \frac{\text{kg} \cdot \text{m}}{\text{sec}^2}}} = 1.1 \text{ sec}$$

The period is 1.1 sec.

4. In this problem, we are given the mass of an object on a spring as well as the period of its motion. Using Equation (11.11), then, we can calculate the spring constant:

$$T = 2 \cdot \pi \sqrt{\frac{m}{k}}$$

$$1.7 \text{ sec} = 2 \cdot \pi \sqrt{\frac{0.400 \text{ kg}}{k}}$$

$$(1.7 \text{ sec})^2 = \left(2 \cdot \pi \sqrt{\frac{0.400 \text{ kg}}{k}}\right)^2$$

$$(1.7 \text{ sec})^2 = 4 \cdot \pi^2 \left(\frac{0.400 \text{ kg}}{k}\right)$$

$$k = 4 \cdot \pi^2 \left(\frac{0.400 \text{ kg}}{(1.7 \text{ sec})^2}\right) = 5.5 \frac{\text{kg}}{\text{sec}^2} = 5.5 \frac{\text{Newtons}}{\text{m}}$$

The spring constant, then, is 5.5 Newtons/m.

5. This problem also uses Equation (11.1). Given both the spring constant and the period, we can calculate the mass:

$$T = 2 \cdot \pi \sqrt{\frac{m}{k}}$$

$$1.1 \text{ sec} = 2 \cdot \pi \sqrt{\frac{m}{30.0 \frac{\text{Newtons}}{\text{m}}}}$$

$$(1.1 \text{ sec})^2 = \left(2 \cdot \pi \sqrt{\frac{m}{30.0 \frac{\text{Newtons}}{\text{m}}}}\right)^2$$

$$(1.1 \text{ sec})^2 = 4 \cdot \pi^2 \left(\frac{m}{30.0 \frac{\text{Newtons}}{\text{m}}}\right)$$

$$m = \frac{(1.1 \text{ sec})^2 \cdot \left(30.0 \frac{\text{kg} \cdot \text{m}}{\text{sec}^2}\right)}{4 \cdot \pi^2} = 0.92 \text{ kg}$$

The mass of the object is <u>0.92 kg</u>.

6. When you work on an object, you are changing its energy. In compressing a spring, you are giving it potential energy. Thus, the amount of work done stretching the spring is equal to the potential energy given to the spring. In the problem, then, the work required to compress the spring will be equal to the spring's potential energy. Given k and x, calculating that is a simple matter:

$$PE = \frac{1}{2} \cdot k \cdot x^2$$

$$PE = \frac{1}{2} \cdot (78.1 \, \frac{\text{Newtons}}{\text{m}}) \cdot (0.511 \, \text{m})^2$$

$$PE = 10.2 \, \text{Newton} \cdot \text{m} = 10.2 \, \text{Joules}$$

It takes <u>10.2 Joules</u> of work, then, to compress the spring 51.1 cm.

7. In order to calculate the maximum speed of an object on a spring, we need to think about where the object reaches its maximum kinetic energy. That's where its speed will also be at a maximum. Well, the object will have maximum kinetic energy at the point where the potential energy is zero. Since we know how the total energy depends on the amplitude of the motion, we can use the fact that potential energy equals zero to calculate the maximum kinetic energy:

$$KE_{spring} + PE_{spring} = \frac{1}{2} \cdot k \cdot A^2$$

$$KE_{max} + 0 = \frac{1}{2} \cdot \left(11.1 \, \frac{\text{Newtons}}{\text{m}}\right) \cdot (0.035 \, \text{m})^2$$

$$KE_{max} = 0.0068 \, \text{Newton} \cdot \text{m} = 0.0068 \, \text{J}$$

Now that we know the maximum kinetic energy, we can calculate the corresponding maximum speed:

$$KE = \frac{1}{2} \cdot m \cdot v^2$$

$$0.0068 \, \text{J} = \frac{1}{2} \cdot (0.034 \, \text{kg}) \cdot v^2$$

$$v = \sqrt{\frac{2 \cdot \left(0.0068 \, \frac{\text{kg} \cdot \text{m}^2}{\text{sec}^2}\right)}{0.034 \, \text{kg}}} = \sqrt{0.40 \, \frac{\text{m}^2}{\text{sec}^2}} = 0.63 \, \frac{\text{m}}{\text{sec}}$$

The maximum speed is <u>0.63</u> m/sec.

8. In order to find the mass's speed, all we need to do is find its kinetic energy. Equation (11.13) gives us a way to relate KE to PE, k, and A. We know A, because we are told how far the spring is stretched before the mass is released. We are given k, so that's taken care of. If we could only determine PE, we would be all set. Well, we can. After all, we are told how far away the mass is from its equilibrium position, and that's x in Equation (11.12). Using this and the spring constant, we can calculate PE:

$$PE = \frac{1}{2} \cdot k \cdot x^2$$

$$PE = \frac{1}{2} \cdot (23.4 \ \frac{\text{Newtons}}{\text{m}}) \cdot (0.050 \ \text{m})^2$$

$$PE = 0.029 \ \text{Newton} \cdot \text{m} = 0.029 \ \text{Joules}$$

Now we can use Equation (11.13):

$$KE_{spring} + PE_{spring} = \frac{1}{2} \cdot k \cdot A^2$$

$$KE_{spring} + 0.029 \ \text{Joules} = \frac{1}{2} \cdot (23.4 \ \frac{\text{Newton}}{\text{m}}) \cdot (0.142 \ \text{m})^2$$

$$KE_{spring} = 0.207 \ \text{Joules}$$

Now that we have the kinetic energy, we can finally get the speed.

$$KE = \frac{1}{2} \cdot m \cdot v^2$$

$$0.207 \ \text{Joules} = \frac{1}{2} \cdot (1.1 \ \text{kg}) \cdot v^2$$

$$v = \sqrt{\frac{2 \cdot 0.207 \ \frac{\text{kg} \cdot \text{m}^2}{\text{sec}^2}}{1.1 \ \text{kg}}} = 0.61 \ \frac{\text{m}}{\text{sec}}$$

5.0 cm from its equilibrium position, then, the mass will be traveling with a speed of <u>0.61 m/sec</u>.

9. This is a simple application of Equation (11.20).

$$T = 2 \cdot \pi \cdot \sqrt{\frac{\ell}{g}}$$

$$T = 2 \cdot \pi \cdot \sqrt{\frac{3.98 \text{ m}}{9.8 \frac{\text{m}}{\text{sec}^2}}} = 4.0 \text{ sec}$$

This pendulum has a period of <u>4.0 sec</u>.

10 In this problem, we are given the period and we know the acceleration due to gravity. Figuring out the length is a simple matter:

$$T = 2 \cdot \pi \cdot \sqrt{\frac{\ell}{g}}$$

$$3.1 \text{ sec} = 2 \cdot \pi \cdot \sqrt{\frac{\ell}{9.8 \frac{\text{m}}{\text{sec}^2}}}$$

$$(3.1 \text{ sec})^2 = 4 \cdot \pi^2 \cdot \left(\frac{\ell}{9.8 \frac{\text{m}}{\text{sec}^2}}\right)$$

$$\ell = \frac{(3.1 \text{ sec})^2 \cdot \left(9.8 \frac{\text{m}}{\text{sec}^2}\right)}{4 \cdot \pi^2} = 2.4 \text{ m}$$

This pendulum, then, has a length of <u>2.4 m</u>.

MODULE 12 - ANSWERS TO THE PRACTICE PROBLEMS

1. This is a simple application of Equation (12.2). Remember, for this equation to work, temperature must be in Celsius.

$$v = (331.5 + 0.60 \cdot T) \, \frac{m}{sec}$$

$$v = (331.5 + 0.60 \cdot 90.0) \, \frac{m}{sec} = 386 \, \frac{m}{sec}$$

Sound waves travel with a speed of <u>386 m/sec</u> in air that is 90 °C.

2. This is a simple application of Equation (12.1). We have to remember, however, that the speed of light is 3.0×10^8 m/sec.

$$f = \frac{v}{\lambda}$$

$$f = \frac{3.0 \times 10^8 \, \frac{m}{sec}}{5.50 \times 10^{-7} \, m} = 5.5 \times 10^{14} \, \frac{1}{sec}$$

This green light, then, has a frequency of <u>5.5×10^{14} Hz</u>.

3. To get the wavelength, we need to know the speed. To get that, we use Equation (12.2):

$$v = (331.5 + 0.60 \cdot T) \, \frac{m}{sec}$$

$$v = (331.5 + 0.60 \cdot 30) \, \frac{m}{sec} = 3.50 \times 10^2 \, \frac{m}{sec}$$

Now we can use Equation (12.1) to answer the question:

$$f = \frac{v}{\lambda}$$

$$1621 \, \frac{1}{sec} = \frac{3.50 \times 10^2 \, \frac{m}{sec}}{\lambda}$$

$$\lambda = \frac{3.50 \times 10^2 \, \frac{m}{sec}}{1621 \, \frac{1}{sec}} = 0.216 \text{ m}$$

The wavelength is <u>0.216 m</u>.

4. In this problem, we are asked to determine the temperature. The only equation we have with temperature in it is Equation (12.2). To use this equation, however, we would need to know the speed. Since we are given both frequency and wavelength, we can get that from Equation (12.1).

$$f = \frac{v}{\lambda}$$

$$545 \, \frac{1}{sec} = \frac{v}{0.651 \, m}$$

$$v = (545 \, \frac{1}{sec}) \cdot (0.651 \text{ m}) = 355 \, \frac{m}{sec}$$

Now that we have the speed, getting the temperature is a snap:

$$v = (331.5 + 0.60 \cdot T) \, \frac{m}{sec}$$

$$355 \, \frac{m}{sec} = (331.5 + 0.60 \cdot T) \, \frac{m}{sec}$$

$$T = \frac{355 - 331.5}{0.60} = 38$$

Since this equation is designed for T to be in Celsius, the temperature is <u>38 °C</u>.

5. We assume that the light reaches the observer instantaneously; thus, the time delay is the time that it takes for the sound to travel the distance between the observer and where the lightning was formed. To determine that distance, we need to determine the speed:

$$v = (331.5 + 0.60 \cdot T) \frac{m}{\sec}$$

$$v = (331.5 + 0.60 \cdot 20.0) \frac{m}{\sec} = 344 \frac{m}{\sec}$$

Now that we know the speed, and we also know that there is no acceleration, we can simply use Equation (3.19) to determine the distance:

$$x = v_o t + \frac{1}{2} a t^2$$

$$x = (344 \frac{m}{\sec}) \cdot (4.2 \sec) + \frac{1}{2} \cdot (0) \cdot (2.5 \sec)^2 = 1.4 \times 10^3 \text{ m}$$

The lightning was formed 1.4 x 10³ m away from the observer.

6. I won't go through all of the steps because you can look at the final diagram and determine what I did in each step. First, we need to determine the relative positions of the object, the radius of curvature, and the mirror. If the radius of curvature is 5.0 cm, then the focal point is 2.5 cm. This puts the object between the focal point and the radius of curvature. Next, we draw our three rays. The first travels from the top of the object horizontally and reflects off the mirror through the focal point. The second starts at the top of the object and travels through the focal point, reflecting horizontally off of the mirror. The last is supposed to travel through the radius and reflect straight back. We can draw this ray *as if* it comes from the radius, and then we can reflect it back to intersect the other two. The result is as follows:

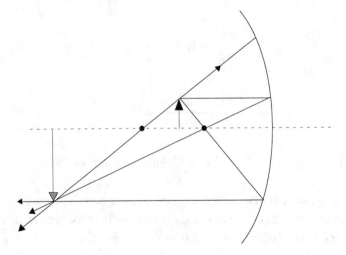

The image is real, inverted, and magnified.

7. This is similar to the problem before, but this time the object is closer to the mirror than the focal point. This means you have to treat the second ray *as if* it came from the focal point and the third ray *as if* it comes from the radius. You will find that the reflected rays do not intersect, so you will have to extrapolate the light rays to find the image.

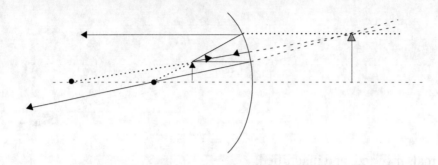

The image is <u>virtual, upright, and magnified</u>.

8. This is a simple application of Equation (12.4), where $n_1 = 1.0$, $\theta_1 = 25°$, and $n_2 = 1.31$.

$$n_1 \cdot \sin\theta_1 = n_2 \cdot \sin\theta_2$$

$$(1.0) \cdot \sin(25) = (1.31) \cdot \sin\theta_2$$

$$\sin\theta_2 = \frac{(1.0) \cdot \sin(25)}{1.31}$$

$$\theta_2 = 19°$$

The angle of refraction is <u>19°</u>.

9. In this problem, the object is much closer to the lens than is the focal point. This means that the second ray will have to be drawn *as if* it came from the focal point. You will find that the light rays do not intersect, so you must extrapolate them backwards to get the image:

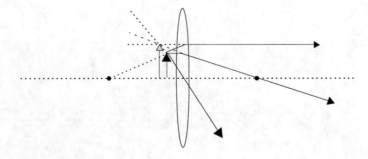

<u>The image is virtual, magnified, and upright</u>.

10. This one is exactly like the first example of a converging lens, just the relative positions of things have changed.

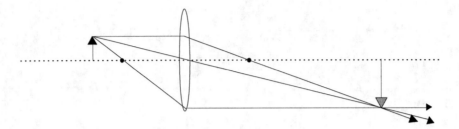

The image is <u>real, inverted, and magnified</u>.

MODULE 13 - ANSWERS TO THE PRACTICE PROBLEMS

1. This is a simple application of Equation (13.1). We have the two charges and the distance between them, so we just plug those values into the equation to calculate the force:

$$F = \frac{k \cdot q_1 \cdot q_2}{r^2}$$

$$F = \frac{(9.0 \times 10^9 \, \frac{\text{Newton} \cdot \cancel{\text{m}^2}}{\cancel{\text{C}^2}}) \cdot (6.7 \, \cancel{\text{C}}) \cdot (3.1 \, \cancel{\text{C}})}{(1.2 \, \cancel{\text{m}})^2} = 1.3 \times 10^{11} \, \text{Newtons}$$

Since one charge is negative and the other positive, this force is attractive. Thus, the <u>charges exert a 1.3 x 10^{11} Newton force towards each other</u>.

2. In this problem, we know the charges and the force and we need to determine the distance between the charges. Equation (13.1) relates these variables. The only thing we have to watch is units. The charges are in mC, but "k" uses C. Thus, we must convert mC to C.

$$F = \frac{k \cdot q_1 \cdot q_2}{r^2}$$

$$2.5 \times 10^5 \, \text{Newtons} = \frac{(9.0 \times 10^9 \, \frac{\text{Newton} \cdot \text{m}^2}{\cancel{\text{C}^2}}) \cdot (3.2 \times 10^{-4} \, \cancel{\text{C}}) \cdot (5.5 \times 10^{-4} \, \cancel{\text{C}})}{(r)^2}$$

$$r^2 = \frac{(9.0 \times 10^9 \, \frac{\cancel{\text{Newton}} \cdot \text{m}^2}{\cancel{\text{C}^2}}) \cdot (3.2 \times 10^{-4} \, \cancel{\text{C}}) \cdot (5.5 \times 10^{-4} \, \cancel{\text{C}})}{2.5 \times 10^5 \, \cancel{\text{Newtons}}} = 0.080 \, \text{m}$$

The distance between the objects, then, is <u>0.080 m</u>.

3. In this problem, we are given everything but the charge of the two objects. You might think at first that there's no way to solve the problem, because there are two things you don't know: both charges. However, the problem tells us that they are the same. So we can call them each "q." That will give us only one variable:

$$F = \frac{k \cdot q_1 \cdot q_2}{r^2}$$

$$2.2 \times 10^5 \text{ Newtons} = \frac{(9.0 \times 10^9 \frac{\text{Newton} \cdot \text{m}^2}{\text{C}^2}) \cdot q \cdot q}{(0.45 \text{ m})^2}$$

$$q = \sqrt{\frac{2.2 \times 10^5 \text{ Newtons} \cdot (0.45 \text{ m})^2}{9.0 \times 10^9 \frac{\text{Newton} \cdot \text{m}^2}{\text{C}^2}}} = 2.2 \times 10^{-3} \text{ C}$$

Each charge has a magnitude of 2.2 mC.

4. The instantaneous electrostatic force on the -3.4 mC particle is zero because the forces exerted by both of the other charges cancel out. The other charges are equal in magnitude and the same distance away from the charge of interest. They each exert an attractive force, pulling the charge of interest in opposite directions. Thus, equal and opposite forces work on the -3.4 mC charge, making the net force zero. At the same time, however, this is not a static system. Even though the net force on the -3.4 mC charge is zero, the net forces acting on the other two charges are not. The -3.4 mC charge attracts each of the other two charges. The other two charges repel each other, but they are farther away from each other than the -3.4 mC charge is from them. As a result, the net charge acting on each of the positive charges is NOT zero. Thus, the other two charges will move towards the -3.4 mC charge.

5. In this problem, we are only worried about the +6.4 mC charge. As a result, we only consider the forces which act on that particular charge. The +1.2 mC charge exerts a repulsive force whose magnitude is:

$$F = \frac{k \cdot q_1 \cdot q_2}{r^2}$$

$$F = \frac{(9.0 \times 10^9 \frac{\text{Newtons} \cdot \text{m}^2}{\text{C}^2}) \cdot (6.4 \times 10^{-3} \text{ C}) \cdot (1.2 \times 10^{-3} \text{ C})}{(1.5 \text{ m})^2} = 3.1 \times 10^4 \text{ Newtons}$$

The other force acting on the +6.4 mC charge is the repulsive force exerted by the +9.1 mC charge.

$$F = \frac{k \cdot q_1 \cdot q_2}{r^2}$$

$$F = \frac{(9.0 \times 10^9 \, \frac{\text{Newtons} \cdot m^2}{C^2}) \cdot (6.4 \times 10^{-3} \, C) \cdot (9.1 \times 10^{-3} \, C)}{(2.0 \, m)^2} = 1.3 \times 10^5 \text{ Newtons}$$

The fact that both forces are repulsive will give us the directions of the force vectors, making our force diagram look like:

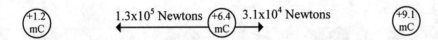

Since the force vectors both point in the same dimension, we can treat them as one-dimensional vectors. This means we can take care of direction with positives and negatives and then simply add the magnitudes together. Using the convention that vectors pointing to the left are negative, the total force is:

$$\mathbf{F}_{total} = -1.3 \times 10^5 \text{ Newtons} + 3.1 \times 10^4 \text{ Newtons} = -1.0 \times 10^5 \text{ Newtons}$$

A negative force means that the final vector points to the left. Thus, the final instantaneous electrostatic force is $\underline{1.0 \times 10^5 \text{ Newtons to the left}}$.

6. Since we are only interested in the -1.2 mC charge, we only need concern ourselves with forces which act on that charge. The +1.2 mC charge exerts an attractive force on it:

$$F = \frac{k \cdot q_1 \cdot q_2}{r^2}$$

$$F = \frac{(9.0 \times 10^9 \, \frac{\text{Newtons} \cdot m^2}{C^2}) \cdot (1.2 \times 10^{-3} \, C) \cdot (1.2 \times 10^{-3} \, C)}{(0.40 \, m)^2} = 8.1 \times 10^4 \text{ Newtons}$$

The +2.4 mC charge also exerts an attractive force on it:

$$F = \frac{k \cdot q_1 \cdot q_2}{r^2}$$

$$F = \frac{(9.0 \times 10^9 \; \frac{\text{Newtons} \cdot \cancel{m^2}}{\cancel{C^2}}) \cdot (1.2 \times 10^{-3} \; \cancel{C}) \cdot (2.4 \times 10^{-3} \; \cancel{C})}{(0.20 \; \cancel{m})^2} = 6.5 \times 10^5 \; \text{Newtons}$$

Our force diagram, then, looks like this:

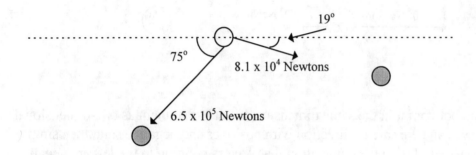

Since these vectors do not point in the same dimension, we will have to add them with trig. First, however, let's define the angles properly. Vector angles are always defined counterclockwise from the +x axis. This means that the first angle is 341°, and the angle for the vector on the left is 255°. Now we can add these vectors:

$$A_x = (8.1 \times 10^4 \; \text{Newtons}) \cdot \cos(341°) = 7.7 \times 10^4 \; \text{Newtons}$$

$$A_y = (8.1 \times 10^4 \; \text{Newtons}) \cdot \sin(341°) = -2.6 \times 10^4 \; \text{Newtons}$$

$$B_x = (6.5 \times 10^5 \; \text{Newtons}) \cdot \cos(255°) = -1.7 \times 10^5 \; \text{Newtons}$$

$$B_y = (6.5 \times 10^5 \; \text{Newtons}) \cdot \sin(255°) = -6.3 \times 10^5 \; \text{Newtons}$$

$$C_x = A_x + B_x = 7.7 \times 10^4 \; \text{Newtons} + -1.7 \times 10^5 \; \text{Newtons} = -9 \times 10^4 \; \text{Newtons}$$

$$C_y = A_y + B_y = -2.6 \times 10^4 \; \text{Newtons} + -6.3 \times 10^5 \; \text{Newtons} = -6.6 \times 10^5 \; \text{Newtons}$$

All that's left to do now is convert these x and y components into vector magnitude and direction:

Magnitude = $\sqrt{C_x^2 + C_y^2} = \sqrt{(-9 \times 10^4 \text{ Newtons})^2 + (-6.6 \times 10^5 \text{ Newtons})^2} = 6.7 \times 10^5$ Newtons

$\theta = \tan^{-1}\left(\dfrac{C_y}{C_x}\right) = \tan^{-1}\left(\dfrac{-6.6 \times 10^5 \text{ Newtons}}{-9 \times 10^4 \text{ Newtons}}\right) = 82°$

Since both the x and y components of the vector are negative, we know that the vector is in quadrant III. This means that we need to add 180.0° to the angle above to properly define the vector angle. The instantaneous electrostatic force on the -1.2 mC charge, then, is <u>6.7 x 10⁵ Newtons at an angle of 262°</u>.

7. When drawing electrical field lines, the lines go out of positive charges and into negative charges. In this case, the negative charge is twice as large as the positive charge, so it has twice as many lines going into it as the positive charge has going out of it. Since our rule of thumb is to draw 8 lines associated with the largest charge, there are 8 lines going into the negative charge. This means there are four lines coming out of the positive charge. Well, since there are fewer lines coming out of the positive charge than what the negative charge can accept, all lines coming from the positive charge end up going into the negative charge. Since these lines are attracted to the negative charge, they want to hit it as soon as possible. Thus, these four lines will enter the negative charge as close as possible to the positive charge. There are, however, still four lines left to draw. They do not come out of the positive charge pictured, but since they are attracted to positive charges, they bend towards it. This makes the picture look as follows:

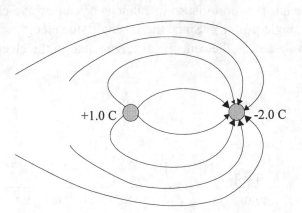

8. <u>If a negative charge were placed in this field, it would travel in the opposite direction that is pointed out by the arrows.</u> Also, since the field strength is greatest where the density of the field lines is greatest, <u>the charge will experience the greatest force when it is near the -2.0 C charge</u>.

9. Since all of the stationary charges have equal magnitudes, they all have 8 lines coming out of them. These lines, however, because they repel each other, bend away form each other when they get close. The result is on the next page.

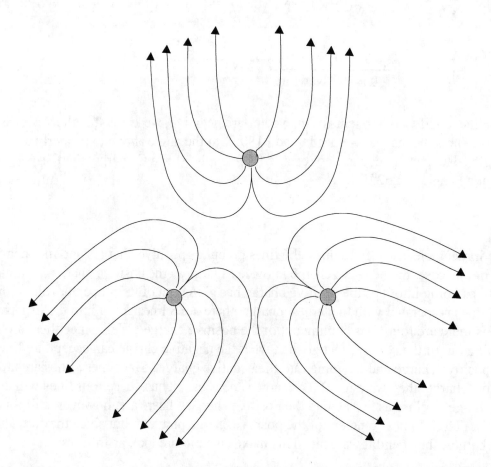

10. The ion in this case has two protons, so its nucleus has a positive charge of 3.2 x 10⁻¹⁹ C (twice the charge of a single proton). Since there is only one electron in the ion, however, we can solve this problem by setting the centripetal force equal to the electrostatic force.

$$\frac{m \cdot v^2}{r} = \frac{k \cdot q_1 \cdot q_2}{r^2}$$

$$v^2 = \frac{k \cdot q_1 \cdot q_2 \cdot \cancel{r}}{r^2 \cdot m} = \frac{k \cdot q_1 \cdot q_2}{r \cdot m} = \frac{(9.0 \times 10^9 \, \frac{\text{Newton} \cdot m^2}{\cancel{C^2}}) \cdot (3.2 \times 10^{-19} \, C) \cdot (1.6 \times 10^{-19} \, C)}{(5.29 \times 10^{-11} \, \cancel{m}) \cdot (9.1 \times 10^{-31} \, kg)}$$

$$v = 3.1 \times 10^6 \, \frac{m}{\text{sec}}$$

This electron has a speed of 3.1 x 10⁶ m/sec.

MODULE 14 - ANSWERS TO THE PRACTICE PROBLEMS

1. This is a simple application of Equation (14.1). You must remember two things, however. First, the units must be correct, so mC must be converted to C. Second, only the stationary charge matters in the potential. The other charge is there simply to confuse you!

$$V = \frac{k \cdot q}{r}$$

$$-4.6 \times 10^7 \text{ V} = \frac{\left(9.0 \times 10^9 \frac{N \cdot m^2}{C^2}\right) \cdot (q)}{0.35 \text{ m}}$$

$$q = \frac{(-4.6 \times 10^7 \frac{N \cdot m}{C}) \cdot (0.35 \text{ m})}{\left(9.0 \times 10^9 \frac{N \cdot m^2}{C^2}\right)} = -1.8 \times 10^{-3} \text{ C}$$

The charge is $\underline{-1.8 \times 10^{-3} \text{ C}}$.

2. This is a simple application of Equation (14.2)

$$PE = q \cdot V$$

$$-12 \text{ J} = (q) \cdot (3.4 \times 10^3 \frac{N \cdot m}{C})$$

$$q = \frac{-12 \text{ J}}{3.4 \times 10^3 \frac{N \cdot m}{C}} = -3.5 \times 10^{-3} \text{ C}$$

The units work here because a J is a N·m. The charge, then, is $\underline{-3.5 \times 10^{-3} \text{ C}}$.

3. To solve this problem, we simply need to calculate the final and initial potential energy and then subtract the two. We will have to convert mC to C in order to do this, however. The initial distance between the charges is 2.20 m and the final distance is 0.20 m, since the freely-moving charge moved 2.0 m closer to the stationary charge. Putting all of this together gives us:

$$V = \frac{k \cdot q}{r}$$

$$V_{initial} = \frac{\left(9.0 \times 10^9 \, \frac{N \cdot m^2}{C^2}\right) \cdot (5.8 \times 10^{-3} \, C)}{2.20 \, m} = 2.4 \times 10^7 \, \frac{N \cdot m}{C} = 2.4 \times 10^7 \, \text{Volts}$$

$$PE = q \cdot V$$

$$PE_{initial} = (-5.1 \times 10^{-3} \, C) \cdot (2.4 \times 10^7 \, \frac{N \cdot m}{C}) = -1.2 \times 10^5 \, N \cdot m = -1.2 \times 10^5 \, J$$

$$V = \frac{k \cdot q}{r}$$

$$V_{final} = \frac{\left(9.0 \times 10^9 \, \frac{N \cdot m^2}{C^2}\right) \cdot (5.8 \times 10^{-3} \, C)}{0.20 \, m} = 2.6 \times 10^8 \, \frac{N \cdot m}{C} = 2.6 \times 10^8 \, \text{Volts}$$

$$PE = q \cdot V$$

$$PE_{final} = (-5.1 \times 10^{-3} \, C) \cdot (2.6 \times 10^8 \, \frac{N \cdot m}{C}) = -1.3 \times 10^6 \, N \cdot m = -1.3 \times 10^6 \, J$$

Change in PE = PE_{final} - $PE_{initial}$ = (-1.3 x 10^6 J) - (-1.2 x 10^5 J) = -1.2 x 10^6 J

Since the change in potential energy is negative, the potential energy <u>decreased by 1.2 x 10^6 J</u>.

4. In solving this problem, we must calculate the initial and final potential energies and then subtract them. In order to get the units to work, however, mC must be converted to C and cm to m.

$$V = \frac{k \cdot q}{r}$$

$$V_{initial} = \frac{\left(9.0 \times 10^9 \, \frac{N \cdot m^2}{C^2}\right) \cdot (-1.8 \times 10^{-3} \, C)}{0.45 \, m} = -3.6 \times 10^7 \, \frac{N \cdot m}{C} = -3.6 \times 10^7 \text{ Volts}$$

$$PE = q \cdot V$$

$$PE_{initial} = (-6.5 \times 10^{-3} \, C) \cdot (-3.6 \times 10^7 \, \frac{N \cdot m}{C}) = 2.3 \times 10^5 \, N \cdot m = 2.3 \times 10^5 \, J$$

$$V = \frac{k \cdot q}{r}$$

$$V_{final} = \frac{\left(9.0 \times 10^9 \, \frac{N \cdot m^2}{C^2}\right) \cdot (-1.8 \times 10^{-3} \, C)}{0.22 \, m} = -7.4 \times 10^7 \, \frac{N \cdot m}{C} = -7.4 \times 10^7 \text{ Volts}$$

$$PE = q \cdot V$$

$$PE_{final} = (-6.5 \times 10^{-3} \, C) \cdot (-7.4 \times 10^7 \, \frac{N \cdot m}{C}) = 4.8 \times 10^5 \, N \cdot m = 4.8 \times 10^5 \, J$$

Change in PE = $PE_{final} - PE_{initial}$ = $(4.8 \times 10^5 \, J) - (2.3 \times 10^5 \, J) = 2.5 \times 10^5 \, J$

Since the change in potential energy is positive, the potential energy <u>increased by 2.5×10^5 J</u>.

5. We solve this problem just like we solved the problems in Module #9. First, we need to calculate the particle's total energy when it starts. At that point, kinetic energy is zero (because it is at rest), and potential energy can be calculated using Equations (14.1) and (14.2):

$$V = \frac{k \cdot q}{r}$$

$$V_{initial} = \frac{\left(9.0 \times 10^9 \, \frac{N \cdot m^2}{C^2}\right) \cdot (4.4 \, C)}{1.4 \, m} = 2.8 \times 10^{10} \, \frac{N \cdot m}{C} = 2.8 \times 10^{10} \, \text{Volts}$$

$$PE = q \cdot V$$

$$PE_{initial} = (1.5 \, C) \cdot (2.8 \times 10^{10} \, \frac{N \cdot m}{C}) = 4.2 \times 10^{10} \, N \cdot m = 4.2 \times 10^{10} \, J$$

Now that we know both the kinetic and potential energy, we know the total energy initially in the system:

$$TE_{initial} = KE_{initial} + PE_{initial} = 0 \, J + 4.2 \times 10^{10} \, J = 4.2 \times 10^{10} \, J$$

That total energy can never change. Thus, if we just calculate the potential energy at the end, we can determine the kinetic energy:

$$V = \frac{k \cdot q}{r}$$

$$V_{final} = \frac{\left(9.0 \times 10^9 \, \frac{N \cdot m^2}{C^2}\right) \cdot (4.4 \, C)}{2.0 \, m} = 2.0 \times 10^{10} \, \frac{N \cdot m}{C} = 2.0 \times 10^{10} \, \text{Volts}$$

$$PE = q \cdot V$$

$$PE_{final} = (1.5 \, C) \cdot (2.0 \times 10^{10} \, \frac{N \cdot m}{C}) = 3.0 \times 10^{10} \, N \cdot m = 3.0 \times 10^{10} \, J$$

Since we know that the total energy still must be 4.2×10^{10} J, we can use this fact and the final potential energy to calculate the final kinetic energy:

$$TE_{final} = KE_{final} + PE_{final}$$

$$4.2 \times 10^{10} \, J = KE_{final} + 3.0 \times 10^{10} \, J$$

$$KE_{final} = 1.2 \times 10^{10} \text{ J}$$

Now we can calculate the speed with Equation (9.3):

$$KE = \frac{1}{2} \cdot m \cdot v^2$$

$$1.2 \times 10^{10} \text{ J} = \frac{1}{2} \cdot (42.3 \text{ kg}) \cdot v^2$$

$$v = 2.4 \times 10^4 \frac{m}{sec}$$

When the particle has traveled to 2.0 m away from the stationary charge, then, its speed is <u>2.4 x 10^4 m/sec</u>.

6. Once again, to solve a problem like this one, we simply look at the energetics involved. First, we need to calculate the particle's total energy when it starts. At that point, kinetic energy is zero (because it is at rest), and potential energy can be calculated using Equations (14.1) and (14.2):

$$V = \frac{k \cdot q}{r}$$

$$V_{initial} = \frac{\left(9.0 \times 10^9 \, \frac{N \cdot m^2}{C^2}\right) \cdot (2.5 \times 10^{-3} \, C)}{0.75 \, m} = 3.0 \times 10^7 \, \frac{N \cdot m}{C} = 3.0 \times 10^7 \text{ Volts}$$

$$PE = q \cdot V$$

$$PE_{initial} = (-9.2 \times 10^{-3} \, C) \cdot (3.0 \times 10^7 \, \frac{N \cdot m}{C}) = -2.8 \times 10^5 \, N \cdot m = -2.8 \times 10^5 \text{ J}$$

Now that we know both the kinetic and potential energy, we know the total energy initially in the system:

$$TE_{initial} = KE_{initial} + PE_{initial} = 0 \text{ J} + -2.8 \times 10^5 \text{ J} = -2.8 \times 10^5 \text{ J}$$

This number can never change. Since the particle moves towards the stationary charge, after traveling 25 cm, the charges will be 50 cm apart. This is the final value for "r."

$$V = \frac{k \cdot q}{r}$$

$$V_{final} = \frac{\left(9.0 \times 10^9 \, \frac{N \cdot m^2}{C^2}\right) \cdot (2.5 \times 10^{-3} \, C)}{0.50 \, m} = 4.5 \times 10^7 \, \frac{N \cdot m}{C} = 4.5 \times 10^7 \, \text{Volts}$$

$$PE = q \cdot V$$

$$PE_{final} = (-9.2 \times 10^{-3} \, C) \cdot (4.5 \times 10^7 \, \frac{N \cdot m}{C}) = -4.1 \times 10^5 \, N \cdot m = -4.1 \times 10^5 \, J$$

Since we know that the total energy still must be -2.8 x 10^5 J, we can use this fact and the final potential energy to calculate the final kinetic energy:

$$TE_{final} = KE_{final} + PE_{final}$$

$$-2.8 \times 10^5 \, J = KE_{final} + -4.1 \times 10^5 \, J$$

$$KE_{final} = 1.3 \times 10^5 \, J$$

We can now use this kinetic energy to solve for the speed.

$$KE = \frac{1}{2} \cdot m \cdot v^2$$

$$1.3 \times 10^5 \, J = \frac{1}{2} \cdot (3.5 \, kg) \cdot v^2$$

$$v^2 = 7.4 \times 10^4 \, \frac{m^2}{sec^2}$$

$$v = 2.7 \times 10^2 \, \frac{m}{sec}$$

When the particle has traveled 25 cm towards the stationary charge, then, its speed is 2.7 x 10^2 m/sec.

7. To solve this problem, we need to find out where the kinetic energy of the particle equals zero. That's the point at which the particle stops and turns around. As always, we start by calculating the total energy from the initial conditions. First, we start with the potential energy:

$$V = \frac{k \cdot q}{r}$$

$$V_{initial} = \frac{\left(9.0 \times 10^9 \; \frac{N \cdot m^2}{C^2}\right) \cdot (1.5 \times 10^{-3} \; C)}{1.2 \; m} = 1.1 \times 10^7 \; \frac{N \cdot m}{C} = 1.1 \times 10^7 \; \text{Volts}$$

$$PE = q \cdot V$$

$$PE_{initial} = (3.8 \times 10^{-3} \; C) \cdot (1.1 \times 10^7 \; \frac{N \cdot m}{C}) = 4.2 \times 10^4 \; N \cdot m = 4.2 \times 10^4 \; J$$

Since the particle is initially moving, it has kinetic energy as well. We need to calculate it in order to get the total energy:

$$KE = \frac{1}{2} \cdot m \cdot v^2$$

$$KE_{initial} = \frac{1}{2} \cdot (5.0 \; kg) \cdot (245 \; \frac{m}{sec})^2$$

$$KE_{initial} = 1.5 \times 10^5 \; J$$

Now that we know both the kinetic and potential energy, we know the total energy initially in the system:

$$TE_{initial} = KE_{initial} + PE_{initial} = 1.5 \times 10^5 \; J + 4.2 \times 10^4 \; J = 1.9 \times 10^5 \; J$$

This number can never change, even when the particle stops. However, when that happens, we know that the particle has zero kinetic energy. Thus, we can calculate the potential energy from these facts:

$$TE_{final} = KE_{final} + PE_{final}$$

$$1.9 \times 10^5 \; J = 0 \; J + PE_{final}$$

$$PE_{final} = 1.9 \times 10^5 \; J$$

In other words, the particle will stop when the potential energy is 1.9×10^5 J. Now we can use Equations (14.1) and (14.2) backwards in order to get the distance at which this occurs. First, since electrical potential is what depends on distance, we need to turn potential energy into electrical potential:

$$PE = q \cdot V$$

$$1.9 \times 10^5 \text{ J} = (3.8 \times 10^{-3} \text{ C}) \cdot V$$

$$V = \frac{1.9 \times 10^5 \text{ J}}{3.8 \times 10^{-3} \text{ C}} = 5.0 \times 10^7 \text{ Volts}$$

The units work out here because a Joule is a C·V. Now that we have electrical potential, we can finally go to Equation (14.1) and calculate the distance between the charges:

$$V = \frac{k \cdot q}{r}$$

$$5.0 \times 10^7 \text{ Volts} = \frac{\left(9.0 \times 10^9 \, \frac{N \cdot m^2}{C^2}\right) \cdot (1.5 \times 10^{-3} \text{ C})}{r}$$

$$r = \frac{\left(9.0 \times 10^9 \, \frac{N \cdot m^2}{\cancel{C^2}}\right) \cdot (1.5 \times 10^{-3} \, \cancel{C})}{5.0 \times 10^7 \, \frac{\cancel{N \cdot m}}{\cancel{C}}} = 0.27 \text{ m}$$

To make the units work out, I replaced Volt with its equivalent unit. So, we finally see that the particle can travel until it is <u>0.27 m away from the charge</u>. At that point, the particle stops, turns around, and heads away from the stationary charge.

8. This is a simple application of Equation (14.3)

$$V = \frac{Q}{C_q}$$

$$V = \frac{0.50 \text{ C}}{3.2 \times 10^{-6} \text{ F}} = \frac{0.50 \, \cancel{C}}{3.2 \times 10^{-6} \, \frac{\cancel{C}}{V}} = 1.6 \times 10^5 \text{ V}$$

The electrical potential of this capacitor is <u>1.6×10^5 Volts</u>.

9. As the proton moves from one plate of the capacitor to the other, it experiences a change in electrical potential. We can use Equation (14.3) to determine the amount of change:

$$V = \frac{Q}{C_q}$$

$$V = \frac{2.2 \times 10^{-3} \text{ C}}{1.4 \times 10^{-6} \text{ F}} = \frac{2.2 \times 10^{-3} \text{ C}}{1.4 \times 10^{-6} \frac{\text{C}}{\text{V}}} = 1.6 \times 10^{3} \text{ V}$$

The fact that the proton moves from the positive plate to the negative one means that this is a decrease in potential, so the change in potential is -1.6 x 10³ Volts.

This change in potential, then, will lead to a change in potential energy, given by Equation (14.2):

$$PE = q \cdot V = (1.6 \times 10^{-19} \text{ C}) \cdot (-1.6 \times 10^{3} \text{ V}) = -2.6 \times 10^{-16} \text{ J}$$

Since the result of the equation is negative, this tells us that the proton's potential energy decreased. Well, if the potential energy of the proton decreased by 2.6×10^{-16} J, then its kinetic energy must have increased by the same amount. Since the kinetic energy started at zero and increased by 2.6×10^{-16} J, we know that the final kinetic energy of the proton is 2.6×10^{-16} J. We can now use Equation (9.3) to determine the final speed:

$$KE = \frac{1}{2} \cdot m \cdot v^2$$

$$2.6 \times 10^{-16} \text{ J} = \frac{1}{2} \cdot (1.7 \times 10^{-27} \text{ kg}) \cdot v^2$$

$$v = 5.5 \times 10^{5} \frac{\text{m}}{\text{sec}}$$

The proton, therefore, moves at the speed of 5.5 x 10⁵ m/sec.

10. What we need to figure out here is the potential. That way, we can determine the charge on the capacitor. Thus, we need to work this problem backwards. Since we know that the electron ends up with a speed of 3.2 x 10⁵ m/sec, we can determine its final kinetic energy:

$$KE = \frac{1}{2} \cdot m \cdot v^2 = \frac{1}{2} \cdot (9.1 \times 10^{-31} \text{ kg}) \cdot (3.2 \times 10^{5} \frac{\text{m}}{\text{sec}})^2 = 4.7 \times 10^{-20} \text{ J}$$

This kinetic energy came from a decrease in the potential energy. Thus, we know that the potential energy decreased by 4.7 x 10^{-20} J. Using Equation (14.2), then, we can determine the change in electrical potential:

$$PE = q \cdot V$$

$$4.7 \times 10^{-20} \text{ J} = (1.6 \times 10^{-19} \text{ C}) \cdot V$$

$$V = \frac{4.7 \times 10^{-20} \text{ J}}{1.6 \times 10^{-19} \text{ C}} = 0.29 \text{ Volts}$$

Now that we know the electrical potential, we can solve for the charge needed to produce such a potential.

$$V = \frac{Q}{C_q}$$

$$0.29 \text{ V} = \frac{Q}{5.1 \times 10^{-6} \text{ F}}$$

$$Q = (0.29 \text{ \cancel{V}}) \cdot (5.1 \times 10^{-6} \frac{C}{\cancel{V}}) = 1.5 \times 10^{-6} \text{ C}$$

In order to achieve the proper speed, then, the physicist must store $\underline{1.5 \times 10^{-6} \text{ C}}$ of charge on the positive plate of the capacitor.

MODULE 15 - ANSWERS TO THE PRACTICE PROBLEMS

1. When you have the voltage and the resistance, the current can be calculated using Equation (15.2):

$$V = I \cdot R$$

$$120 \text{ V} = I \cdot (501 \text{ } \Omega)$$

$$I = \frac{120 \text{ V}}{501 \text{ } \Omega} = 0.24 \text{ A}$$

The current is <u>0.24 A</u> and flows as follows:

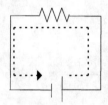

2. This is another application of Equation (15.2).

$$V = I \cdot R$$

$$9.0 \text{ V} = (0.50 \text{ A}) \cdot (R)$$

$$R = \frac{9.0 \text{ V}}{0.50 \text{ A}} = 18 \text{ } \Omega$$

The resistance is <u>18 Ω</u>.

3. This problem gives power and voltage and asks for current. Equation (15.3) relates these quantities:

$$P = I \cdot V$$

$$315 \text{ W} = I \cdot (120 \text{ V})$$

$$I = \frac{315 \text{ W}}{120 \text{ V}} = 2.6 \frac{\text{W}}{\text{V}} = 2.6 \text{ A}$$

If you think about our discussion of units, you will see that a Watt per Volt is an Amp. The current, then, is 2.6 A.

4. In this problem, we have current and resistance and need to determine power. Equation (15.4) relates these quantities.

$$P = I^2 \cdot R = (1.2 \text{ A})^2 \cdot (15 \text{ }\Omega) = 22 \text{ Watts}$$

The circuit uses 22 Watts of power.

5. In this circuit, the resistors are hooked up in parallel, because the current has a choice as to which resistor to travel through. As a result, we use Equation (15.6) to determine the effective resistance:

$$\frac{1}{R_{effective}} = \frac{1}{R_1} + \frac{1}{R_2}$$

$$\frac{1}{R_{effective}} = \frac{1}{151 \text{ }\Omega} + \frac{1}{215 \text{ }\Omega} = 0.00662 \frac{1}{\Omega} + 0.00465 \frac{1}{\Omega}$$

$$\frac{1}{R_{effective}} = 0.01127 \frac{1}{\Omega}$$

$$R_{effective} = 88.73 \text{ }\Omega$$

The effective resistance, then, is 88.73 Ω.

6. These resistors are hooked up in series because all of the current must go through both of them. Thus, Equation (15.5) gives us the effective resistance:

$$R_{effective} = R_1 + R_2 = 151 \text{ }\Omega + 215 \text{ }\Omega = 366 \text{ }\Omega$$

The effective resistance is 366 Ω.

7. To determine the power drawn by the circuit, we must determine the current drawn by the entire circuit. To do that, we need the resistance of the entire circuit to use in Equation (15.2). Thus, we need the effective resistance of all three resistors. Since they are hooked up in series, the effective resistance is given by Equation (15.5):

$$R_{effective} = R_1 + R_2 + R_3 = 20.0 \text{ }\Omega + 9.0 \text{ }\Omega + 12.0 \text{ }\Omega = 41.0 \text{ }\Omega$$

Now that we have the effective resistance, we can use Equation (15.2) to get the current:

$$V = I \cdot R$$

$$9.0 \text{ V} = I \cdot (41.0 \text{ } \Omega)$$

$$I = \frac{9.0 \text{ V}}{41.0 \text{ } \Omega} = 0.22 \text{ A}$$

This current, then, can be used in Equation (15.3) to calculate the power drawn by the circuit:

$$P = I \cdot V = (0.22 \text{ A}) \cdot (9.0 \text{ V}) = 2.0 \text{ Watts}$$

The circuit draws 2.0 Watts of power.

8. Once again, we need to get the effective resistance in order to get the current in order to get the power. Since the resistors are in parallel here, we need Equation (15.6) to get the effective resistance:

$$\frac{1}{R_{effective}} = \frac{1}{R_1} + \frac{1}{R_2} + \frac{1}{R_3}$$

$$\frac{1}{R_{effective}} = \frac{1}{9.0 \text{ } \Omega} + \frac{1}{19.0 \text{ } \Omega} + \frac{1}{15.0 \text{ } \Omega} = 0.11 \frac{1}{\Omega} + 0.0526 \frac{1}{\Omega} + 0.0667 \frac{1}{\Omega}$$

$$\frac{1}{R_{effective}} = 0.23 \frac{1}{\Omega}$$

$$R_{effective} = 4.3 \text{ } \Omega$$

We can use this resistance in Equation (15.2) to get the current:

$$V = I \cdot R$$

$$120 \text{ V} = I \cdot (4.3 \text{ } \Omega)$$

$$I = \frac{120 \text{ V}}{4.3 \text{ } \Omega} = 28 \text{ A}$$

This current, then, can be used in Equation (15.3) to calculate the power drawn by the circuit:

$$P = I \cdot V = (28 \text{ A}) \cdot (120 \text{ V}) = 3.4 \times 10^3 \text{ Watts}$$

The power draw is 3.4×10^3 Watts.

9. This is a more complicated circuit, so we first need to find all of the resistors in parallel and reduce them to their effective resistors. As the current moves from the positive side of the battery, it passes through the fuse and then gets to make a choice between the 201 Ω and the 301 Ω resistor. This means that those two resistors are in parallel. The 211 Ω is in series because all current must travel through it. Thus, we first must reduce the two parallel resistors. Equation (15.6), which I assume you can use by now, tells us that the effective resistance is 120 Ω. This reduces the circuit to a series circuit.

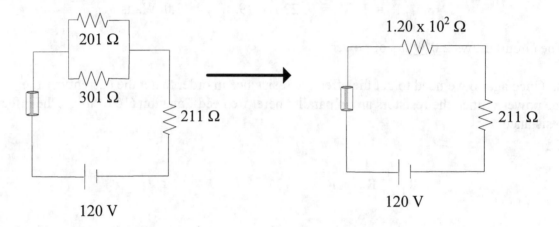

Now we can use Equation (15.5) to calculate the effective resistance of these two resistors in series. That gives us 331 Ω as the effective resistance of the circuit. Now we can calculate the current with Equation (15.2):

$$V = I \cdot R$$

$$120 \text{ V} = I \cdot (331 \text{ Ω})$$

$$I = \frac{120 \text{ V}}{331 \text{ Ω}} = 0.36 \text{ A}$$

Since the current is only 0.36 A, the circuit should have the 1 Amp fuse.

10. To get the power we need the current. To get that, we need the effective resistance. As the current leaves the positive end of the battery, it must all travel through the 31 Ω resistor, so that is a series resistor. The current does have a choice, however, between the 27 Ω resistor and the 21 Ω resistor. This means that they are in parallel. Equation (15.6) tells us that the effective resistance of these two resistors is 12 Ω. The current then has another choice between the 15 Ω resistor and the 11 Ω resistor. This means they are in parallel, and Equation (15.6) tells us that their effective resistance is 6.3 Ω. The circuit, then, can be simplified to:

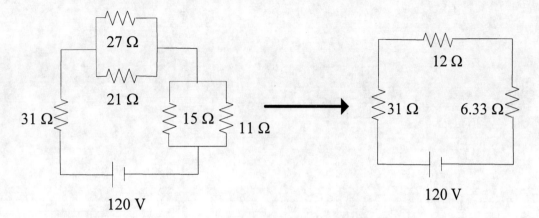

Now we just have a series circuit. Equation (15.5) tells us that the effective resistance of this circuit is 49 Ω. Now we can calculate the current:

$$V = I \cdot R$$

$$120 \text{ V} = I \cdot (49 \text{ Ω})$$

$$I = \frac{120 \text{ V}}{49 \text{ Ω}} = 2.4 \text{ A}$$

This current, then, can be used in Equation (15.3) to calculate the power drawn by the circuit:

$$P = I \cdot V = (2.4 \text{ A}) \cdot (120 \text{ V}) = 2.9 \times 10^2 \text{ Watts}$$

The circuit draws $\underline{2.9 \times 10^2 \text{ Watts}}$ of power.

Tests

TEST FOR MODULE #1

Be sure to write all your answers with the proper number of significant figures and to list the units that go with your answers!

1. What are the meanings for the metric prefixes "milli," "centi," and "kilo?"

2. Which is heavier, a 0.3 g rock or a 30.0 mg rock?

3. The following numbers are the results of several measurements of a football field (which is supposed to be 100.0 yards long):

   a. 113.1 yards
   b. 1.0×10^2 yards
   c. 99.126 yards

Which of these three numbers represents the most precise measurement? Which is the most accurate?

4. If we observe a fishing bobber float on the surface of the water, what can we conclude about the density of the bobber compared to the density of the water?

5. What is the volume of the liquid in the following graduated cylinder?

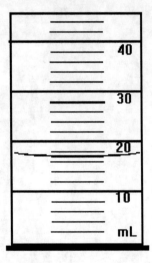

6. If the mass of the liquid in problem #5 was 32.13 grams, what would its density be?

7. How many cm are in 16.2 m?

8. If an object has a mass of 345.6 mg, what is its mass in grams?

9. If a football field is 100.0 yards long, how many miles long is it? (1 yard = 3.000 feet, 1 mile = 5.280 x 10^3 feet).

10. The density of gold is 19.3 grams per mL. A miner finds a gold-colored nugget whose volume is 34.2 mL and whose mass is 661 grams. Is it really a nugget of gold?

11. A fisherman wants to determine the volume of his lead sinker. If lead's density is 11.4 grams per cc and the sinker has a mass of 0.123 kg, what is the volume of the sinker?

12. Convert the number 3478 to scientific notation.

13. Convert the number 1.245 x 10^{-4} into decimal form.

14. The volume of a sphere is given by the equation

$$V = \frac{4}{3}\pi r^3$$

Where π = 3.1415, and r is the radius of the sphere. If a sphere's radius is 3.1 m, what is its volume in liters?

15. The mass of an object is measured on earth and also on the moon. How do the two measured masses compare?

TEST FOR MODULE #2

1. The acceleration of a car is zero. Does this mean that its velocity is also zero?

2. If the velocity of a plane is reported as 100 m/sec and its acceleration is reported as -10 m/sec^2, is the plane speeding up or slowing down?

3. What is the difference between speed and velocity?

4. A physicist measures the instantaneous velocity of an object in three different places along its journey. The velocities were 2 m/sec, 4 m/sec, and 8 m/sec. She then measures the object's average velocity throughout the journey. Which of the following values is most likely to be that average velocity: 1 m/sec, 3 m/sec, or 16 m/sec?

5. A person is walking in an airport at a constant velocity of 1.3 m/sec. What is the person's acceleration?

6. A football player receives a kickoff and runs 98 yards straight into the end zone. After spiking the ball and doing his dance, he turns around and runs 31 yards back to his teammates for some serious celebrating. How much distance did he travel? What was his final displacement?

7. A car is traveling down a one-lane country road at 21 miles/hour. Up ahead, a truck is traveling the opposite direction at 15 miles/hour. If they are 0.20 miles apart, how long will it be before they run into each other, assuming neither one of them slows down?

8. A car starts at rest and accelerates as quickly as possible. If the acceleration is 1.2 m/sec^2 and the car accelerates for 10.0 seconds, what will the car's final velocity be?

(Test continues on the next page.)

Questions 9 - 12 refer to the figure below.

A car's motion is described by the following displacement versus time graph:

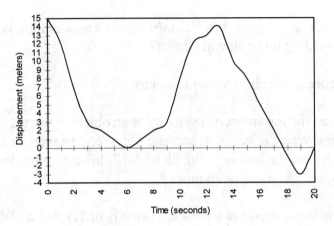

9. What is the car's velocity at 6.0 seconds?

10. Is the car's velocity larger at 11.2 seconds or 18.0 seconds?

11. What is the car's velocity at 17.0 seconds?

12. How many times does the car change direction?

Questions 13 - 15 refer to the figure below.

Consider an object whose motion is given by the following velocity versus time graph:

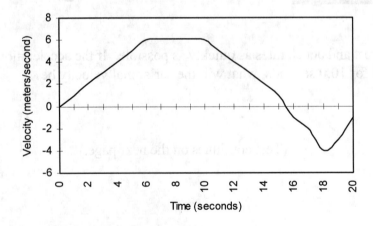

13. What is the object's acceleration at 2.0 seconds?

14. What is the object's acceleration at 8.0 seconds?

15. Over what time intervals does the object speed up?

TEST FOR MODULE #3

1. A car's engine is giving out. The car erratically speeds up and slows down. Can we use the equations we derived in this module to analyze its motion? Explain.

2. Suppose you are doing physics experiments outside in a thick fog. Fog makes the molecules and atoms in the air more difficult to move. In those circumstances, would you expect the air resistance in your experiments to be more, less, or the same as it would be without fog?

3. Under the circumstances described in problem #2, would you expect the terminal velocity of the objects you are studying to be lower, higher, or the same as without fog?

4. A bullet is shot out of a gun straight up in the air. When it falls back to earth, a radar gun measures its velocity to be 125 m/sec downwards when it reaches the same place from which it was fired. What was the velocity of the bullet when it left the gun?

5. When the bullet in problem #4 reached its maximum height, what was its velocity?

6. When the bullet in problem #4 reached its maximum height, what was its acceleration?

7. When an object moves at constant velocity, what is its acceleration?

8. If you drop two objects and they do not fall at the same rate, what can you conclude about the effect of air resistance on the two objects?

9. A boat travels for 4.0 hours at a constant velocity of 10.0 m/sec. How far does the boat travel? (1 hr = 3600 sec)

10. To measure the height of a building without a ruler or tape measure, an engineer drops a rock off the top of the building and finds out that it takes 4.9 seconds for the rock to reach the ground. How high is the building?

11. A car accelerates from rest at 0.21 m/sec^2 for 2.0 minutes. What is its final velocity?

12. A driver is driving down the road and sees a deer in his headlights. The driver hits the brakes, which provide a deceleration of -7.0 m/sec^2. If the car's initial velocity was 25 m/sec, how far will the car skid before coming to a stop?

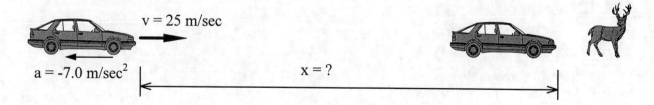

13. If a bicyclist starts from rest and accelerates at 1.1 ft/sec² for 1.0 minute, how far will the bicyclist travel?

14. A cannon shoots its cannonball straight up in the air with an initial velocity of 2.0×10^2 m/sec. What is the maximum height that the cannonball will reach?

15. A happy graduate is throwing her graduation cap into the air. She releases the cap 6.0 feet from the ground and it takes 2.1 seconds to land on the ground. What initial velocity did she give to the cap?

TEST FOR MODULE #4

For Problems 1 - 7, consider the three vectors below:

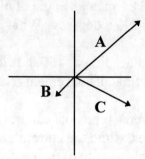

1. Which has the largest magnitude?

2. Which has the largest angle?

3. If an airplane is traveling southeast, which vector would be most likely to represent its velocity?

4. If **A** is the acceleration of an object and **B** is its velocity, would the object be speeding up or slowing down?

5. Graphically add vectors **A** and **C**.

6. Draw the vector that would represent **-C**.

7. Graphically subtract vector **C** from vector **B**.

8. Why do we graphically add and subtract vectors when we can analytically do so?

9. An airplane is flying in a northeasterly wind as shown below. Draw the vector that represents the velocity which the pilot should maintain so that the airplane can arrive at the airport due east of its present location:

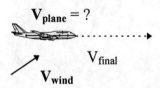   

10. Vector **A** has a magnitude of 3.2 m and a direction of 15 degrees. What are its x- and y-components?

11. The velocity vector of a bike has an x-component of 1.7 m/sec and a y-component of -1.1 m/sec. What are the magnitude and direction of the velocity vector?

12. Vector **A** has a magnitude of 31.1 m at an angle of 60.0 degrees, and vector **B** has a magnitude of 11.4 m at an angle of 290.0 degrees. What is the sum of these two vectors?

13. A hiker travels 1.8 miles with a heading of 150.0 degrees and then 3.2 miles at a heading of 350.0 degrees. What is the scout's final displacement relative to his starting point?

14. A boat travels across a wide river. If the boat can travel at a speed of 15 mph and its pilot heads in a direction of 130.0 degrees, while the current's velocity is 2.1 mph at a heading of 200.0 degrees, what will be the final velocity of the boat?

15. A plane heads due north ($\theta = 90.0°$) at a speed of 200.0 mph. If the wind's velocity is southeast ($315°$) at 15.0 mph, what will be the actual velocity of the plane?

TEST FOR MODULE #5

1. A projectile launched from the ground hits a target on the ground several meters away from the launch site. If the projectile hits the target with a speed of 124 m/sec, what was the initial speed with which the projectile was fired?

2. A bullet is fired at a target which is at the same height as the gun from which the bullet was fired. If the bullet takes 0.44 seconds to hit the target, how long did it take for the bullet to reach its maximum height?

3. A military rocket is equipped with thruster engines that allow it to navigate towards its target. These thrusters fight gravity so that the vertical acceleration the projectile experiences is greatly reduced. In addition, the thrusters allow the rocket to accelerate horizontally. Do you expect this rocket to travel on a parabolic path? Why or why not?

4. What one condition must apply before you can use the range equation [Equation (5.9)]?

5. If you double the initial speed of a projectile, what happens to that projectile's range?

6. Suppose you used Equation (5.9) to calculate the range of a projectile. If you were to make another calculation in which you included the effect of air resistance, would the calculated range increase or decrease?

7. The y-component of a projectile's velocity is 12.1 m/sec. When the projectile once again passes by the height from which it was launched, what is the y-component of its velocity?

8. Which of the following objects would move in a parabolic path?
    a. A rock thrown with an angle of 40° relative to the ground
    b. A jet airplane in flight
    c. A rocket in space

9. A projectile is fired with a certain initial velocity from the surface of the earth. If the same projectile were taken to the moon and shot from the surface of the moon with the same initial velocity, would it have a larger, smaller, or identical range? Why?

10. An explorer follows these directions:

    Travel with a constant velocity of 2.1 m/sec at a heading of 120.0 degrees for 1.1 hours, then turn to a heading of 200.0 degrees and travel at the same speed for 2.3 hours.

What will the explorer's final displacement be?

11. An archer lies on the ground and lets loose an arrow with an initial velocity of 112 ft/sec at $\theta = 40.0°$. What is the maximum height that the arrow will reach?

12. What is the range of a gun that fires its bullets with a velocity of 175 m/sec at an angle of 30.0 degrees?

13. Aimed at an angle of 45 degrees, a cannon has a range of 4,123 feet. What is the initial speed of a cannonball launched from this cannon?

14. A diver is on a hill overlooking a lake. If the hill is 45 feet above the surface of the water and the diver jumps horizontally ($\theta = 0.000°$) with an initial speed of 20.0 ft/sec, how far from the hill will the diver hit the lake?

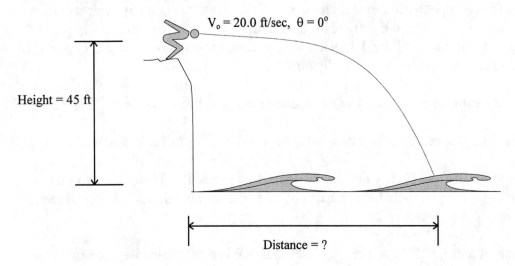

15. In an archery competition, an archer stands on a hill so that the arrow in her bow is 20.0 meters above the ground and is aimed at an angle of 0.000 degrees relative to the ground. If the archer is trying to hit a target that lies on the ground 118 m away, with what initial velocity must she launch her arrow? (NOTE: The picture is not drawn to scale!)

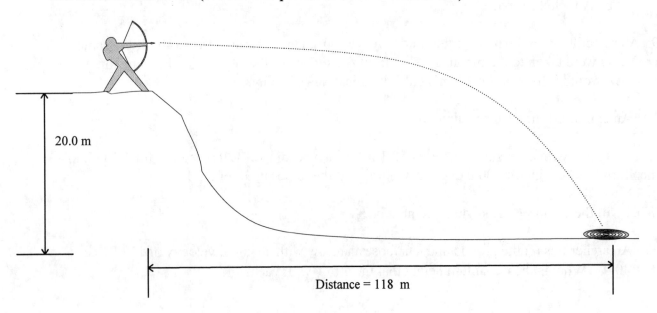

TEST FOR MODULE #6

1. State Newton's three laws of motion in your own words.

2. A bicyclist is traveling at high speed and is not watching where he is going. His front tire catches in a sewer grating, causing the bike to stop almost instantly. This makes the rider fall off of the bike. Which way will the biker fall and why?

3. Two objects have different weights. The first is twice as heavy as the second. If they are each placed on the same surface, which will experience the greatest friction? How much greater will the frictional force be?

4. You find a measurement of 34 kg written in a laboratory notebook. What physical characteristic is this a measurement of?

5. From experience, we all know that you must push an object harder to get it moving than you do to keep it moving once it has already started. Explain, from an atomic scale point of view, why this is the case.

6. A physicist is using tools that measure length in cm, mass in mg, and time in minutes. If the physicist measures a force and does not convert any units, what will the force unit be?

7. A parent asks a physics student to move a couch. The physics student, having just read about Newton's Third Law, has just learned that for every force applied, there is an equal and opposite force. He therefore reasons that when he pushes the couch, an equal and opposite force will cancel out his force, making it impossible to move the couch. What is wrong with the student's reasoning?

8. A pole-vaulter vaults over a 10-foot high bar and lands on a large cushion. When the pole-vaulter hits the cushion, he exerts a force on it. What is the evidence of this force? What is the equal and opposite force demanded by Newton's Third Law, and what is the evidence of that force?

9. Assuming that there is no friction, how much force is necessary to accelerate a 152 gram object at 1.21 m/sec^2?

10. An object has a mass of 34 slugs. What is its weight?

11. A rocket ship weighs 24,561 Newtons on earth. How much will it weight on Jupiter? (The acceleration due to gravity on Jupiter is 23.2 m/sec^2.)

12. If a rock weighs 1,231 pounds on Venus, how much will it weigh on earth? (The acceleration due to gravity on Venus is 28 ft/sec^2.)

13. What is the frictional force that keeps a 567-pound statue from moving on its platform (μ_s = 0.32, μ_k = 0.20)?

14. A 721-kg car accelerates on a road whose coefficients of friction are μ_s = 0.42, μ_k = 0.30. If the engine is capable of supplying a 3,123 Newton force, how fast will the car accelerate, provided that it is already moving?

F = 3,123 Newtons

15. A mover pushes a 1,451-pound piano across a cement driveway (μ_s = 0.42, μ_k = 0.24). Assuming that the mover has already gotten the piano moving, how much force must the mover push with in order to keep the piano moving at a constant velocity?

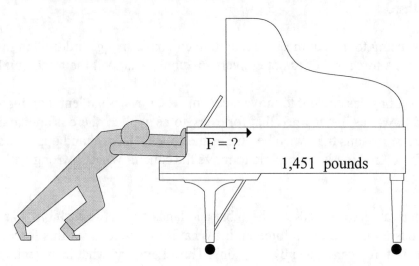

F = ?

1,451 pounds

TEST FOR MODULE #7

1. An object moves with constant velocity. Is it in a state of equilibrium? If so, what kind of equilibrium?

2. Suppose you throw a Frisbee in the air towards your friend. It spins at a constant rate, but the speed with which it moves towards your friend is changing due to the wind. Is the Frisbee in any state of equilibrium? If so, what kind?

3. What are the conditions required for an object to be in static rotational equilibrium?

4. Two plumbers are working at a site. The first tries to unscrew a pipe with a wrench and cannot budge it. The second one is twice as strong as the first one. He also has a wrench that is twice as long. How much more torque can the second plumber exert compared to the first?

5. A driver is trying to eat with one hand and steer with the other (not a good thing to do). If the driver is using only one hand, which of the two situations below will give the driver the most power in turning the wheel?

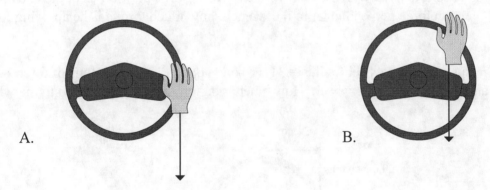

A.   B.

6. Which of the following situations will result in the most tension in the two strings?

A.   B.

7. Although Newton·meter is the standard unit for torque, there are other possible units. Which of the units below is a possible unit for torque?

slug·foot, pound·inch, kg·km, Newton·kg

8. A 53.4-Newton child sits in a swing as pictured on the next page. What is the tension in the two ropes of the swing?

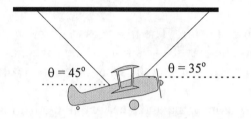

9. A 9.0-kg model airplane is tied to the ceiling with two strings as shown below. What is the tension in each string?

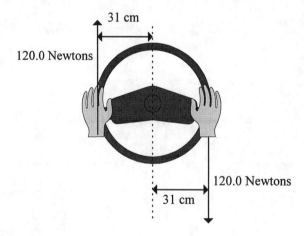

10. A mechanic unscrews a nut with a 15-cm wrench. If he grasps the wrench at the end and applies a 456-Newton force perpendicular to the wrench, how much torque is he applying to the nut?

11. A driver grasps a steering wheel (radius = 31 cm) and applies 120.0 Newtons of force with each hand in an attempt to turn the wheel. How much total torque is being applied to the wheel?

12. Two plumbers are trying to unscrew a pipe. The first can exert a force of 345 Newtons while the second is capable of exerting a force of 545 Newtons. If the first has a wrench that is 45 cm long and the second has a wrench that is 25 cm long, which will be most likely to succeed?

13. Two children (35 kg and 41 kg) balance on a see-saw as pictured below. Where must the boy sit to achieve balance?

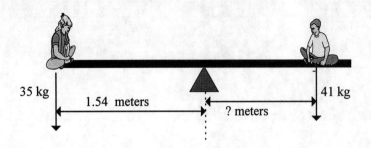

14. A 45-kg block slides down an incline that is angled at 41 degrees. If the coefficient of kinetic friction between the block and the incline is 0.45, what is the block's acceleration?

15. A child drags two boxes (112 kg and 95.1 kg) behind her bicycle as shown below. If she pulls the boxes with a force of 123 Newtons, what is the tension in the string that connects the boxes? Ignore friction in this problem.

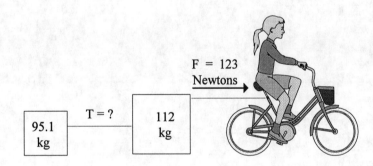

TEST FOR MODULE #8

($G = 6.67 \times 10^{-11} \frac{\text{Newton} \cdot \text{m}^2}{\text{kg}^2}$, Mass of sun = 2.0×10^{30} kg, Mass of earth = 5.98×10^{24} kg)

1. Several physical characteristics are listed below. Which do not change in uniform circular motion?

a. *velocity*   b. *speed*   c. *centripetal acceleration*   d. *magnitude of the centripetal acceleration*
e. *centripetal force*   f. *magnitude of the centripetal force*

2. An object moves in a circle with constant speed as pictured below. Draw in the velocity, acceleration, and force vectors.

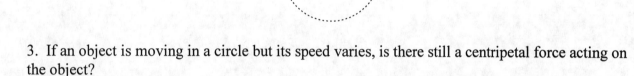

3. If an object is moving in a circle but its speed varies, is there still a centripetal force acting on the object?

4. If an object is moving in a circle at a constant speed and suddenly its speed decreases by a factor of two, how must the centripetal acceleration change to keep the object moving in the same circle?

5. If two massive objects were placed near each other on a frictionless surface, what would happen?

6. In a physics experiment, two objects were placed a certain distance apart and their gravitational attraction was measured. Afterwards, the distance between them was decreased by a factor of 4. What happened to the gravitational force between them when the distance was changed?

7. Why do the planets orbit the sun in nearly circular orbits?

8. Two satellites of equal mass orbit the earth. The speed of the first satellite is 30% greater than the speed of the second. Which orbits the earth at the higher altitude?

9. During a test drive, a car is driven at a constant speed in a circle of radius 31.0 meters. If the car is traveling at 25.0 m/sec, what centripetal acceleration is required to keep the car traveling in the circle?

10. In the problem above, what is the period and frequency of the car's motion?

11. A car is turning on a curve whose radius of curvature is 23.1 meters. If the car is traveling at 13 m/sec, what is the minimum coefficient of kinetic friction necessary to keep the car on the road?

12. Two objects of equal mass exert a gravitational force of 12.1 Newtons on each other when they are 15 cm apart. Determine the mass of each object.

13. Venus orbits the sun at a distance of 1.1×10^{11} meters. How fast does it travel in its orbit?

14. The rings of Saturn are actually composed of small rocks, presumably from a moon that was crushed by a catastrophic event. If the orbital radius of one of those rocks is 6.1×10^7 meters, what is the rock's orbital period? (mass of Saturn = 5.7×10^{26} kg)

15. A satellite orbits the earth with a speed of 9.4×10^2 m/sec. What is its orbital period?

TEST FOR MODULE #9

1. Distinguish between work and energy. What do they have in common? What are their differences?

2. Define kinetic and potential energy.

3. In a typical electrical power plant, a fuel (like coal) is burned to boil water. The steam from that water is used to turn large fans which operate the electrical generator. The turning of the fans causes the generator to make electricity, which moves down the wires. Starting with the fuel, describe each of the energy transformations that take place. Tell what form (chemical, electrical, etc.) and type (potential or kinetic) of energy is transformed in each stage of this process.

4. A man tries to pull an object as pictured below. How could he change the way he is pulling in order to increase the work that he can perform?

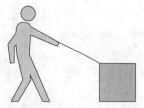

5. Is it ever possible for the work done by friction to *increase* the kinetic energy of an object?

6. A roller coaster is designed so that the car is pulled up to the top of the first hill by a chain and pulley system. After that, however, the car coasts the rest of the way through the track, never being pulled or pushed again until its brakes stop it. With these conditions, what is wrong with the roller coaster designed below?

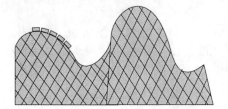

7. Two workers must pick up bricks that lie on the ground and place them on a worktable. They each pick up the same number of bricks and put them on the same height worktables. The first gets the job done in one-half the time that the second takes. Did one of the workers do more work than the other? If so, which one? Did one of the workers exert more power than the other? If so, which one?

8. When friction takes energy from an object, where does the energy go?

9. A refrigerator repairman pushes a refrigerator 75 cm to get access to the panel behind it. If the worker did 175 Joules of work, what force did he exert?

10. A toy car (m = 25 grams) on a track starts from rest at the top of a 2.3 meter hill and rolls down. The car then goes up a ramp and flies off the track (see diagram below). If the end of the ramp is 0.91 meters high, at what speed does the car leave the track?

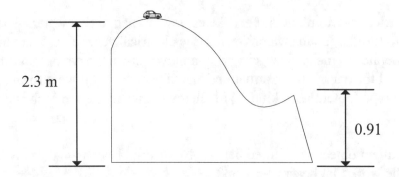

11. A bicyclist travels up a 14.1-m hill and reaches the top with a speed of 9.7 m/sec. At that point, the bicyclist stops pedaling and coasts down the hill and up the next one. If the next hill is 2.3 meters high, how fast will the bicyclist be traveling when it reaches the top of the second hill? The combined mass of the bicycle and the cyclist is 89 kg.

12. An angry carpenter kicks a 2.3-kg block of wood across the floor with an initial velocity of 5.1 m/sec. If the coefficient of kinetic friction between the wood and the floor is 0.67, how far will it travel before coming to rest?

13. A 345-gram ball rolls down a 3.1-m hill. If it started from rest and gets to the bottom with a speed of 4.2 m/sec, how much work did friction perform?

14. A 101 Watt light bulb burns for 30.0 minutes. How much energy was consumed?

15. A machine must move an object 15.1 meters by exerting a 121 Newton force. If the machine must get the job done in 2.3 minutes, how much power must the machine's motor provide?

TEST FOR MODULE #10

1. Which of the following vector pairs are possible velocity and momentum vectors for the same object?

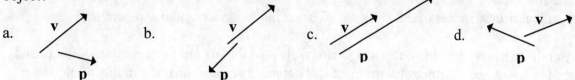

2. Two men run down a road. They each run at the same velocity, but the first has more momentum than the second. How is that possible?

3. A physics teacher tells you that she observed two objects of equal mass traveling with identical speeds. She claims that their momenta were different, however. How is that possible?

4. Explain (in terms of the concepts we have learned in this module) how air bags reduce injuries in traffic accidents.

5. When a balloon is blown up and then released, it flies about as the air escapes. Why?

6. Which of the following are legitimate units for angular momentum?

$$\frac{kg \cdot m}{sec^2}, \frac{g \cdot mm^2}{min}, \frac{g \cdot m}{sec^2}, \frac{in^2 \cdot slug}{hr}$$

7. Can a cat with no tail always land on its feet when it falls? Why or why not?

8. A 125-gram toy car travels with a velocity of 5.4 m/sec 300 degrees southeast. What is its momentum?

9. A 365-kg car traveling horizontally at 21.1 m/sec slams into a tree and comes to a halt in 0.22 seconds. What force did the tree exert on the car in order to stop it?

10. A 250.0-gram baseball is thrown horizontally to a batter at 38 m/sec. The batter hits the ball and sends a line drive horizontally at -52 m/sec. If the bat delivered a force of -201 Newtons, how long were the ball and bat in contact?

11. A 2.34-kg gun has a recoil velocity of 5.2 m/sec. At what velocity does it fire its 95-gram bullets?

12. A 60.0-kg ice skater stands motionless and catches an 8.0-kg medicine ball (a heavy ball used in athletic training and physical therapy) that was traveling towards her horizontally at 3.7 m/sec. What will her velocity be after she catches the ball?

13. A 975-kg train car coasts slowly (v = 3.1 m/sec) under a hopper that fills it with coal. If the car's velocity slows to 1.2 m/sec, what mass of coal was loaded into it?

14. The moon has a mass of 7.36 x 10^{22} kg and orbits the earth with a linear speed of 1.00 x 10^3 m/sec and an orbital radius of 3.8 x 10^8 m. What is the moon's angular momentum?

15. A child twirls a toy plane on a string. He twirls it so that its radius of motion is 45 cm and its speed is 3.9 m/sec. Without twirling, he lets out more string so that the radius of motion increases to 98 cm. What is the plane's new speed?

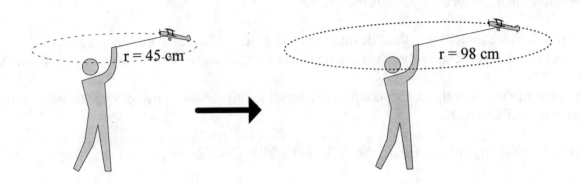

TEST FOR MODULE #11

1. A system is pulled away from its equilibrium position and then released. If "z" is a measurement of the system's displacement from equilibrium, "C" is a constant, and **F** represents the restoring force, which of the following relationships between the variables will result in simple harmonic motion?

   a. $\mathbf{F} = -C \cdot z^2$ b. $\mathbf{F} = -C \cdot z$ c. $\mathbf{F} = -\dfrac{C}{z}$

2. A mass/spring system is pulled 5.4 cm from equilibrium and then released. Its period of motion is measured to be 1.1 seconds. The system is then brought to a halt. Next, it is displaced 10.0 cm from its equilibrium position and released again. What will its new period of motion be?

3. In a mass/spring system, where does the mass have the greatest amount of kinetic energy?

4. In a mass/spring system, where does the mass experience the greatest acceleration?

5. You are doing an experiment to measure the period of simple harmonic motion exhibited by a pendulum. The instructions tell you to displace the pendulum from its equilibrium position and release it. Unfortunately, the instructions do not tell you how far from equilibrium to displace the pendulum. Keeping in mind that you are trying to study simple harmonic motion, should you displace the pendulum a large or small distance from equilibrium?

6. You construct two pendulums. The first has twice the mass as the second. Otherwise, they are identical. How do the periods of the pendulums compare?

7. You construct two pendulums. The first is significantly shorter than the second. Otherwise, they are identical. How do the periods of the pendulums compare?

8. If the motion of a mass/spring system has an amplitude of 12.1 cm, how far from equilibrium was it initially displaced?

9. A 34.5 kg mass bounces on a spring (k = 12.1 Newtons/m). What is the period of its motion?

10. An object is hung on a spring, causing the spring to stretch 15.1 cm. If the force constant of the spring is 2.9 Newtons/meter, what is the mass of the object?

11. A 5.61-kg fish is hung on a spring scale. The spring stretches 9.2 cm in response. The fisherman, in an effort to make the fish seem heavier, pulls down on the fish, displacing it an additional 3.1 cm. His wife, trying to make him honest, slaps his hand, making him release the fish. As a result, the fish begins bouncing up and down on the spring scale. What is the period of its motion?

Questions 12 - 14 refer to the following situation:

A 45.0 kg mass is hung on a spring (k = 5.11 Newtons/meter). The system is then displaced 30.0 cm from equilibrium and released.

12. What is the total energy of the mass/spring system?

13. What is the mass's maximum speed?

14. What will its speed be when it is 10.0 cm from equilibrium?

15. An astronaut has lost her star charts and has no idea what planet she has just landed on. Although she has no star charts, she does have a list of all known planets and their gravitational accelerations. Being a good physicist, she puts together a 0.75-m long pendulum and measures its period to be 2.8 seconds. Given the list below, what planet is she on?

Planet	Acceleration Due To Gravity (m/sec^2)
Uranus	9.5
Venus	8.7
Mars	3.8
Neptune	14
Pluto	5.2

TEST FOR MODULE #12

1. A wave oscillates in the horizontal dimension and propagates in the vertical dimension. Is it a longitudinal or a transverse wave?

2. What kind of images are formed in a flat mirror: real or virtual?

3. Substance A is more dense than substance B. In which substance does light travel faster?

4. Given substances A and B above, in which does sound travel faster?

5. A soprano sings with a very high pitch while a tenor sings with a lower pitch. Which singer's sound waves have the larger wavelength?

6. If substance A has half the index of refraction of substance B, which is more dense?

7. State the Law of Reflection.

8. Two singers sing at exactly the same pitch, but the first is louder than the second. What is the difference between the two singers' sound waves?

9. What is the wavelength of light if its frequency is 1.2×10^3 Hz?

10. What is the frequency of a sound wave traveling in 25 °C air if its wavelength is 0.512 m?

11. What is the temperature if a 612-Hz sound wave has a wavelength of 0.580 m?

12. If the temperature during a thunderstorm is 11 °C and the thunderclap is heard 2.9 seconds after the lightning flash is observed, how far away from the observer was the lightning?

13. An object is placed 15.0 cm away from a mirror whose radius of curvature is 10.0 cm. Draw a ray tracing diagram to illustrate what the image will look like. Is it real or virtual? Is it upright or inverted? Is the image magnified, reduced, or essentially the same size as the object?

14. A converging lens has a focal point that is 12.0 cm from its center. If an observer looks through the lens at an object 6.0 cm from the lens, what will the image look like? Is it real or virtual? Is it upright or inverted? Is the image magnified, reduced, or essentially the same size as the object?

15. An object is placed 5.0 cm away from a mirror whose radius of curvature is 20.0 cm. Draw a ray tracing diagram to illustrate what the image will look like. Is it real or virtual? Is it upright or inverted? Is the image magnified, reduced, or essentially the same size as the object?

TEST FOR MODULE #13

$$(k = 9.0 \times 10^9 \frac{Newtons \cdot m^2}{C^2})$$

1. A physicist uses a positively charged rod to give a metal sphere a negative charge. Did the physicist charge by conduction or induction?

2. An ion has more protons than electrons. What is the sign of its electrical charge?

3. Two charged objects (one positive and one negative) are held apart with a long bar:

If the charge of these two objects does not change, even after a long time, is the bar made of an insulator or a conductor?

4. When we use Coulomb's Law to calculate the force that exists between two charges which are free to move, we say that we are calculating the instantaneous electrostatic force. Why must we say "instantaneous."?

5. Two charged objects are isolated in space. The charge on the positive object is twice the charge on the negative object. If the positive object exerts a 5.0 N force on the negative object, how much force does the negative object exert on the positive one?

6. Two charged objects are held in place and the electrostatic force between them is measured. If, suddenly, half of the charge is taken out of one of the objects, what happens to the magnitude of the force?

7. Two charged objects are held in place and the electrostatic force between them is measured. If, suddenly, the distance between the objects is halved, what happens to the magnitude of the force?

8. A +2.1 C charge is placed 45 cm from a -9.3 C charge. What is the force between them? Is it attractive or repulsive?

9. Two charged objects that are 1.2 m apart experience a repulsive instantaneous electrostatic force of 4.3×10^5 Newtons. If one of the objects is charged at +3.1 mC, what is the charge (include the sign) of the other object?

10. Three charges are arranged as follows:

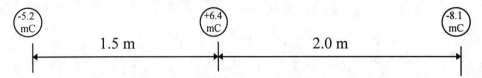

What is the instantaneous electrostatic force on the +6.4 mC charge?

11. Three charges are arranged as follows:

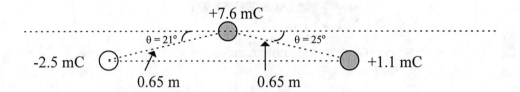

What is the instantaneous electrostatic force on the +7.6 mC particle?

12. Consider the following electrostatic field:

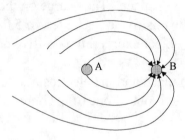

a. Is particle B positively or negatively charged?

b. If the magnitude of particle A's charge is 1.0 C, what is the magnitude of particle B's charge?

c. Where should a freely-moving charged be placed so that it experiences the *least* acceleration?

13. Four protons are placed at the vertices of a square. Draw the electrical field that results:

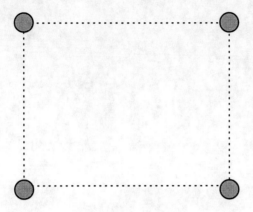

14. What is the speed of an electron (q = -1.6 x 10^{-19} C, m = 9.1 x 10^{-31} kg) in a Li^{2+} ion if it orbits the nucleus at a distance of 5.9 x 10^{-12} m? This ion is made up of three protons (q = 1.6 x 10^{-19} C each), 2 neutrons, and 1 electron.

TEST FOR MODULE #14

$$(k = 9.0 \times 10^9 \; \frac{N \cdot m^2}{C^2})$$

1. A physicist's notebook is full of experimental observations. An astute student says that there is a problem with the following entry:

    "Electrical Potential in Trial 1 = 1.2 J"

    What is wrong with this entry?

2. A negative particle is placed near a stationary charge. If the electrical potential that the particle experiences is negative, is the stationary charge positive or negative?

3. A particle is placed near a stationary positive charge. If the particle's potential energy is positive, is the particle positive or negative?

4. A negative charge moves towards a stationary positive charge. As the two charges get closer together, does the potential energy increase or decrease?

5. What is the definition of a capacitor?

6. If a negative particle moves from the positive plate of a capacitor to the negative plate, does it experience an increase or decrease in electrical potential?

Problems 7 and 8 refer to the diagram below:

```
            - | -
  ......electron beam......▶
            + | +
```

7. Draw the path of the electron beam, assuming that the charge stored in the capacitor is large.

8. Draw the path of the electron beam, assuming that the charge stored in the capacitor is small.

9. A television repair person notices that when a television is turned on, a single, bright, vertical line appears at the center of the screen. The line goes from the very top of the screen to the very bottom, while the rest of the screen stays black. The repairman immediately decides that one of the controlling capacitors inside the cathode ray tube has ceased to function. Which capacitor does the repair person suspect and why?

10. A +2.3 mC charge is placed 35 cm away from a stationary -3.4 mC charge. What is the electrical potential?

11. In a 15 Volt potential, a particle has a potential energy of -0.45 J. What is the particle's charge?

12. A -1.5 mC charge is moving towards a stationary +6.8 mC charge. What is the change in potential energy as the particle moves from 75 cm away from the stationary charge to only 33 cm from the stationary charge? Did the potential energy increase or decrease?

13. A +1.5 mC charged particle (m = 4.3 kg) is placed 1.2 m from a +4.4 mC stationary charge. If it starts from rest, how fast will the particle be traveling when it is 2.0 m away from the stationary charge?

14. A +2.7 mC charged particle (m = 1.5 kg) is shot with an initial velocity of 202 m/sec towards a +1.8 mC stationary charge. If the particle starts out 1.4 meters from the stationary charge, how close will it come to the charge before turning around and moving away?

15. A proton (m = 1.7 x 10^{-27} kg, q = +1.6 x 10^{-19} C) is placed at edge of the positive plate of a 2.4 x 10^{-6} F capacitor. If the capacitor holds 17 mC of charge on its positive plate, how fast will the proton be moving when it reaches the negative plate of the capacitor?

TEST FOR MODULE #15

1. If current is really the flow of electrons through a conductor, why do circuit diagrams always show current as flowing from the positive side of the battery to the negative side?

2. A string of lights that do not seem to work suddenly all light up when a single bulb is replaced. Is this a series or parallel circuit?

3. Why do all conductors have at least some resistance in them?

4. How does an electrical heater produce heat?

5. Draw the current that flows in the following diagram:

6. In which of the circuits below will the light bulb actually light?

a.   b.

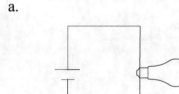 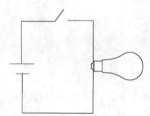

7. A circuit has a single resistor of 56 Ω and it draws 2.3 A of current. What is the voltage of its power source?

8. A motor has a resistance of 4.1 Ω and draws 1.7 A of current. What is the voltage of the battery it runs on?

9. If a resistor uses 98 Watts of power and draws 2.1 A of current, what is its resistance?

10. Determine the effective resistance of the following circuit:

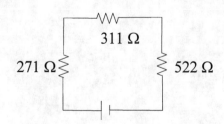

11. Determine the effective resistance of the following circuit

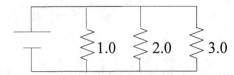

12. What is the power drawn by the following circuit?

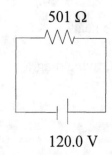

13. An electrician has 1 Amp, 3 Amp, and 5 Amp fuses. Which should be used for this circuit?

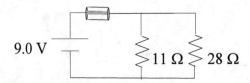

14. What is the power drawn by the following circuit?

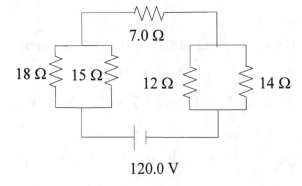

15. An electrician has 1 Amp, 3 Amp, and 5 Amp fuses. Which should be used for this circuit?

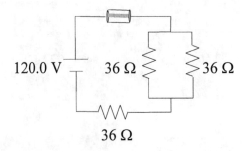

TEST FOR MODULE #16

1. A magnet that is free to move is placed in the following magnetic field at the point labeled "X." In which direction does the magnet's south pole point?

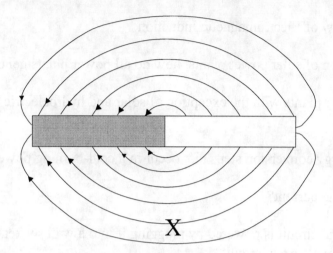

2. Draw the magnetic field lines for the situation below:

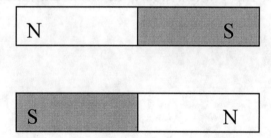

3. What is wrong with the following picture?

4. A substance is attracted to a magnet, but no matter how hard you try, you cannot turn it into a magnet. Is the substance diamagnetic, paramagnetic, or ferromagnetic?

5. A substance is made from atoms that are not magnetic. Is the substance diamagnetic, paramagnetic, or ferromagnetic?

6. A substance is made from atoms that are magnetic, but these atoms do not arrange themselves in magnetic domains. Is the substance diamagnetic, paramagnetic, or ferromagnetic?

7. How can you turn a ferromagnetic substance that is not a magnet into a magnet?

8. Draw the magnetic field lines for the following current-carrying wire.

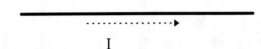
I

9. State Faraday's Law of Electromagnetic Induction.

10. With the exception of solar power plants, how do all power plants generate electricity?

11. What do all power plants with the exception of solar and hydroelectric power plants use to turn their turbines?

12. Describe the energy conversion steps that occur in a coal-burning power plant.

13. What is alternating current?

14. A certain electrical circuit is powered by plugging it into a wall socket. Is this an alternating current circuit or a direct current circuit?

15. Why is a household outlet rated for 120 V when, in fact, the maximum voltage that it delivers is 170 V?

Solutions To The

Tests

ANSWERS TO THE MODULE #1 TEST

1. The prefix "milli" means 0.001. The prefix "centi" means 0.01. The prefix "kilo" means 1,000.

2. To properly compare these measurements, we need to get them into the same units. I will convert mg to g, although you could just as easily convert g to mg.

$$\frac{30.0 \text{ mg}}{1} \times \frac{0.001 \text{ g}}{1 \text{ mg}} = 0.0300 \text{ g}$$

Since 30.0 mg is really equal to 0.0300 g, then it is the smallest of the two numbers. Thus, the 0.3 gram rock is the heaviest.

3. Remember, precision is determined by how many decimal places there are. The more decimal places, the more precise the number. Accuracy, on the other hand, tells us how close to the true value a measurement is. Since a football field is supposed to be 100.0 yards long, then (c) is the most precise measurement, but (b) is the most accurate.

4. If the bobber floats on the surface of the water, its density must be less than that of water.

5. It takes 5 dashes to go from 10 to 20 mL. This must mean that each dash is worth 2 mL. Since the meniscus is between the third and forth dash, then the answer is somewhere between 16 and 18 mL. The meniscus looks just slightly above halfway between the two, so we could estimate that it is a little more than 17 mL. Since we are always to estimate one more decimal place than the scale reads, we could say that the volume is 17.1 mL (any number between 16.8 and 17.4 would be fine)

6. Density is mass divided by volume. The volume is provided in the answer to question #5, so we take 32.13 g and divide by the volume given in #5 (17.1 mL), and we get 1.88 grams per mL. (If the answer in #5 is different, this answer will be different. Check it by taking the mass and dividing by the answer the student gave for #5. There should only be three significant figures in any answer to this question.)

7. $\dfrac{16.2 \text{ m}}{1} \times \dfrac{1 \text{ cm}}{0.01 \text{ m}} = 1.62 \times 10^3 \text{ cm}$

The answer is 1.62×10^3 cm.

8. $\dfrac{345.6 \text{ mg}}{1} \times \dfrac{0.001 \text{ g}}{1 \text{ mg}} = 0.3456 \text{ g}$ Thus, the answer is 0.3456 grams.

9. To do this problem, you cannot directly convert yards to miles, because you have no direct relationship between the two. Thus, you must first convert feet. Then you can convert to miles.

Exploring Creation With Physics Solutions and Tests (Test Solutions): 159

$$\frac{100.0 \text{ yards}}{1} \times \frac{3.000 \text{ feet}}{1 \text{ yard}} = 300.0 \text{ feet}$$

$$\frac{300.0 \text{ feet}}{1} \times \frac{1 \text{ mile}}{5.280 \times 10^3 \text{ feet}} = 0.05682 \text{ miles}$$

The answer is 0.05682 miles.

10. To test whether or not this nugget is gold, we simply compute the density. If the density is 19.3 g/mL, then the nugget is gold. If not, the nugget is not gold.

$$\rho = \frac{m}{V} = \frac{661 \text{ g}}{34.2 \text{ mL}} = 19.3 \frac{\text{g}}{\text{mL}}$$

The nugget is gold!

11. The density is given in g/cc. Now remember, this is the same as g/mL, but we will go ahead and use cc's. Since density is mass divided by volume, if we know density and mass, we can get volume. Notice, however, that there are conflicting units. The density uses grams (g/cc), but the mass is given in kg. Thus, before I solve the problem, I need to convert from kg to g.

$$\frac{0.123 \text{ kg}}{1} \times \frac{1000 \text{ g}}{1 \text{ kg}} = 123 \text{ g}$$

Now that all units agree, we can use the equation for density to solve for volume.

$$\rho = \frac{m}{V}$$

$$11.4 \frac{\text{g}}{\text{cc}} = \frac{123 \text{ g}}{V}$$

$$V = \frac{123 \text{ g}}{11.4 \frac{\text{g}}{\text{cc}}} = 10.8 \text{ cc}$$

The volume, then, is 10.8 cc's.

12. To get the decimal point to the right of the first digit, we need to move it 3 spaces to the left. Thus, the exponent on the ten will be a 3. Since this number is large, the exponent is positive. Thus, the answer is 3.478×10^3.

13. There is an exponent of 4 on the ten. This means that we must move the decimal 4 places. Since the exponent is negative, we need to move the decimal so that the number is less than one. Thus, the answer is <u>0.0001245</u>.

14. This is the hardest problem on the test. First, we need to plug the radius in the equation and get the volume:

$$V = \frac{4}{3} \cdot (3.1415) \cdot (3.1 \text{ m})^3 = 1.2 \times 10^2 \text{ m}^3$$

Although this **is** the volume of the sphere, it is not the answer, because the question asked for the volume in LITERS. Do we know of a way to convert from m^3 to liters? No, because we don't know of a relationship between them. We do know, however, that a mL is the same as a cm^3. Thus, we can convert from m^3 to cm^3, which then is the same as mL. Once we have mL, we can get to liters:

$$\frac{1.2 \times 10^2 \text{ m}^3}{1} \times \left(\frac{1 \text{ cm}}{0.01 \text{ m}}\right)^3$$

$$\frac{1.2 \times 10^2 \text{ m}^3}{1} \times \frac{1 \text{ cm}^3}{0.000001 \text{ m}^3} = 1.2 \times 10^8 \text{ cm}^3 = 1.2 \times 10^8 \text{ mL}$$

$$\frac{1.2 \times 10^8 \text{ mL}}{1} \times \frac{0.001 \text{ liter}}{1 \text{ mL}} = 1.2 \times 10^5 \text{ liters}$$

The volume is <u>1.2×10^5 L</u>.

15. <u>The mass would be the same.</u> Unlike weight, mass never changes, regardless of where you are.

ANSWERS TO THE MODULE #2 TEST

1. <u>No</u>. A car moving at constant velocity has zero acceleration but not zero velocity.

2. Since velocity and acceleration have opposite signs, we can conclude that <u>the plane is slowing down</u>.

3. <u>Speed is not a vector quantity and velocity is</u>. This means that speed has no directional information, while velocity does.

4. <u>3 m/sec</u>. The average velocity should be somewhere in between the instantaneous velocities, like in Experiment 2.2.

5. <u>0</u>, because the velocity is not changing.

6. Distance: Direction is irrelevant here, so we just add up the total distance. He went 98 yards and then 31 yards more, for a total of <u>129 yards</u>. Displacement: Direction is important here. If we define motion from where he caught the ball to the endzone as positive, he traveled 98 yards followed by -31 yards. The final displacement is the sum of the two, or <u>67 yards from where he caught the ball, towards the end zone</u>.

7. The collision time will depend on their *relative* velocity. If we define rightward motion as positive, the car is traveling at 21 mph and the truck travels at -15 mph. The relative velocity is the difference, or 21 mph - (-15 mph) = 36 mph. Now we can just use the definition of velocity to get the time interval:

$$v = \frac{\Delta x}{\Delta t}$$

$$36 \frac{mi}{hr} = \frac{0.20 \, mi}{\Delta t}$$

$$\Delta t = \frac{0.20 \, \cancel{mi}}{36 \frac{\cancel{mi}}{hr}} = 0.0056 \, hr$$

They will collide in <u>0.0056 hours</u> or 20 seconds.

8. This is a simple application of the definition of acceleration:

$$a = \frac{\Delta v}{\Delta t}$$

$$1.2 \frac{m}{sec^2} = \frac{\Delta v}{10.0 \text{ sec}}$$

$$\Delta v = (1.2 \frac{m}{sec^2}) \cdot (10.0 \text{ sec}) = 12 \frac{m}{sec}$$

Since the car started from rest, the change in velocity will also be the final velocity, so the car's final velocity is 12 m/sec.

9. At 6.0 seconds, the slope of the graph is flat. This means that the slope is zero, and therefore the velocity is 0.

10. The slope tells us velocity. The steeper the slope, then, the larger the velocity. At 11.2 seconds, the curve is nearly flat. At 18.0 seconds, however, the slope is steeply falling. Thus, the velocity is greatest at 18.0 seconds.

11. From 15.0 sec to 19.0 seconds, the curve looks like a straight line. Over that interval, then, the slope (and therefore the velocity) is constant. If we calculate the slope of the line during that interval, we have the velocity at any time in that interval. At 15.0 seconds, the displacement looks to be about 9.0 m. At 19.0 seconds, the displacement is about -3.0 m. Thus, the slope is:

$$\text{slope} = \frac{9.0 \text{ m} - (-3.0 \text{ m})}{15.0 \text{sec} - 19.0 \text{ sec}} = -3.0 \frac{m}{sec}$$

The velocity is -3.0 m/sec ****

12. The sign of the slope is the sign of the velocity, which tells us the direction of travel. Thus, each time the slope changes sign, it corresponds to a change in direction. The slope stays negative until after 6.0 second, where it turns positive. This is the first direction change. It changes from positive to negative at 13.0 seconds, and it changes from negative back to positive at 19.0 seconds. Thus, the direction changed 3 times.

13. From 0.0 sec to 6.0 seconds, the curve looks like a straight line. Over that interval, then, the slope (and therefore the acceleration) is constant. If we calculate the slope of the line during that interval, we have the acceleration at any time in that interval.

$$\text{slope} = \frac{6.0 \frac{m}{sec} - 0.0 \frac{m}{s}}{6.0 \text{sec} - 0.0 \text{ sec}} = 1.0 \frac{m}{sec^2}$$

The acceleration is $\underline{1.0 \text{ m/sec}^2}$. ****

14. The curve is flat from 6.0 seconds to 10.0 seconds. Over that whole range, then the acceleration is $\underline{0}$.

15. Objects speed up when their acceleration vectors and their velocity vectors point in the same direction. On the graph, this means that velocity and slope (acceleration) must have the same signs. Velocity is positive from 0.0 to 6.0 seconds. During that time, the slope is positive as well. This means that the object is speeding up. From 6.0 to 10.0 seconds, there is no acceleration, because the slope is zero. Thus, it is not speeding up. From 10.0 seconds to 15.5 seconds, the velocity is positive but the slope is negative. In that region, then, it is slowing down. From 15.5 seconds to 18.0 seconds, both the velocity and the slope are negative, so it is speeding up again. Finally, after 18.0 seconds, the velocity is still negative but the slope is positive. Thus, the object is slowing down during that interval. The object speeds up then, over two intervals: $\underline{0.0 - 6.0 \text{ seconds, } 15.5 - 18.0 \text{ seconds}}$ ****

****Your student's answers may vary a bit, since they come from reading a graph.

ANSWERS TO THE MODULE #3 TEST

1. No, the car's acceleration is not constant.

2. Air resistance should be greater, because it results from the falling object needing to push the molecules in the air out of the way. If it is harder to do that, then the air resistance will be greater.

3. Since air resistance is large, then it will be hard for things to free fall. Thus, terminal velocity will be lower.

4. When an object is thrown up in the air, it will travel upwards until gravitational acceleration rids it of its upward velocity. When it begins to fall, gravitational acceleration will speed it up again, giving it all of its velocity back by the time it reaches the height at which it was launched. At that point, if it is traveling down at 125 m/sec, then it must have initially been traveling 125 m/sec up.

5. At the maximum height, a projectile's vertical velocity is 0.

6. At any point in free fall, an object's acceleration is 9.8 m/sec² down or 32 ft/sec² down.

7. Constant velocity means no change in velocity. Therefore, acceleration must be 0.

8. Since all objects would fall at the same rate if it weren't for air resistance, one of the objects must be more affected by air resistance than the other.

9. Since the boat is moving at a constant velocity, its acceleration is zero. Since the velocity is constant, we also know that we can use it as either the initial or final velocity, or both. So, we have initial and/or final velocity, acceleration, and time, and we want to determine distance. Equation (3.19) will work:

$$x = v_o t + \frac{1}{2} a t^2$$

Before we use the equation, however, we need to make our units work. Right now, time is in hours while velocity is in m/sec. To fix this, we will convert hours to seconds:

$$\frac{4.0 \text{ hr}}{1} \times \frac{3600 \text{ sec}}{1 \text{ hr}} = 1.4 \times 10^4 \text{ sec}$$

Now that our units agree, we can solve the equation:

$$\mathbf{x} = (10.0 \frac{m}{sec}) \cdot 1.4 \times 10^4 \text{ sec} + \frac{1}{2} \cdot 0 \cdot (1.4 \times 10^4 \text{ sec})$$

$$\mathbf{x} = 1.4 \times 10^5 \text{ m}$$

The boat travels $\underline{1.4 \times 10^5 \text{ m}}$.

10. In this problem, we are given the time it takes a rock to fall. If we can calculate its displacement in that time, we will be able to determine the height of the building. Since the rock is in free fall, we know its acceleration is 9.8 m/sec². We also know that since it was dropped, its initial velocity is zero. I went ahead and left the acceleration positive, indicating that down is the positive direction. If we have initial velocity, acceleration, and time, and we want to determine displacement, Equation (3.19) is the one to use:

$$\mathbf{x} = \mathbf{v}_o t + \frac{1}{2}\mathbf{a}t^2$$

$$\mathbf{x} = (0) \cdot (4.9 \text{ sec}) + \frac{1}{2} \cdot (9.8 \frac{m}{sec^2}) \cdot (4.9 \text{ sec})^2$$

$$\mathbf{x} = 1.2 \times 10^2 \text{ m}$$

The building is $\underline{1.2 \times 10^2 \text{ meters}}$ or $\underline{3.8 \times 10^2 \text{ feet}}$ (if you used english units) tall.

11. This is a simple application of Equation (3.6), as long as we remember to convert minutes to seconds to get our units in agreement:

$$\frac{2.0 \text{ min}}{1} \times \frac{60 \text{ sec}}{1 \text{ min}} = 1.2 \times 10^2 \text{ sec}$$

$$\mathbf{v} = \mathbf{v}_o + \mathbf{a}t$$

$$\mathbf{v} = 0 + (0.21 \frac{m}{sec^2}) \cdot (1.2 \times 10^2 \text{ sec}) = 25 \frac{m}{sec}$$

The final velocity is $\underline{25 \text{ m/sec}}$.

12. This problem give us initial velocity (23 m/sec), acceleration (-7.0 m/sec²) and final velocity (0). From that, we need to calculate displacement. Equation (3.15) will work for this.

$$v^2 = v_o^2 + 2ax$$

$$(0)^2 = (25\frac{m}{sec})^2 + 2\cdot(-7.0\frac{m}{sec^2})\cdot(x)$$

$$x = \frac{(25\frac{m}{sec})^2}{-14.0\frac{m}{sec^2}} = 45\text{ m}$$

The stopping distance is <u>45 meters</u>.

13. This problem uses Equation (3.19), but we first need to get units consistent:

$$\frac{1.0\ \cancel{min}}{1} \times \frac{60\text{ sec}}{1\ \cancel{min}} = 6.0\times 10^1\text{ sec}$$

$$x = v_o t + \frac{1}{2}at^2$$

$$x = (0)\cdot(6.0\times 10^1\text{ sec}) + \frac{1}{2}\cdot(1.1\frac{ft}{sec^2})\cdot(6.0\times 10^1\text{ sec})^2$$

$$x = 2.0\times 10^3\text{ ft}$$

The cyclist travels <u>2.0×10^3 ft</u>.

14. The cannonball will reach its maximum height when its velocity is zero. We know final velocity, initial velocity, and acceleration, and we want to determine distance. Equation (3.15) will help us with that.

$$v^2 = v_o^2 + 2ax$$

$$(0)^2 = (2.0\times 10^2\frac{m}{sec})^2 + 2\cdot(-9.8\frac{m}{sec^2})\cdot(x)$$

$$x = \frac{-4.0\times 10^4\ \frac{m^2}{sec^2}}{2\cdot(-9.8\frac{m}{sec^2})} = 2.0\times 10^3\text{ m}$$

The cannonball reaches a maximum height of 2.0 x 10³ meters.

15. The place students trip up here is in defining the directions and sticking to it. The cap is in free fall. If we define downward motion as negative, then its acceleration is -32 ft/sec². We also know the time (2.1 seconds). Most students trip up here. The displacement is -6.0 ft. Why negative? It landed 6.0 ft *lower* than where it started. Its final displacement, then, is 6.0 ft *down*, which, according to our definition, means -6.0 ft. Now we can use Equation (3.19):

$$x = v_o \cdot t + \frac{1}{2} \cdot a \cdot t^2$$

$$-6.0 \text{ ft} = (v_o) \cdot (2.1 \text{ sec}) + \frac{1}{2} \cdot (-32 \frac{\text{ft}}{\text{sec}^2}) \cdot (2.1 \text{ sec})^2$$

$$v_o = \frac{-6.0 \text{ ft} - \frac{1}{2} \cdot (-32 \frac{\text{ft}}{\text{sec}^2}) \cdot (2.1 \text{ sec})^2}{2.1 \text{ sec}} = 31 \frac{\text{ft}}{\text{sec}}$$

The graduate originally threw the cap up with a velocity of 31 ft/sec.

ANSWERS TO THE MODULE #4 TEST

1. The vector with the largest magnitude is represented by the longest arrow, thus, **A**.

2. Vector angle is defined counterclockwise from the +x axis, thus **C**.

3. South is considered below the x-axis and east is considered to be right of the y-axis, thus **C**.

4. Under these conditions, the velocity and acceleration vectors are pointing in opposite directions. This means that it is <u>slowing down</u>.

5.

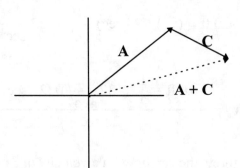

6.

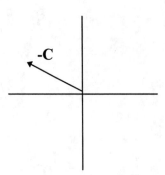

7.

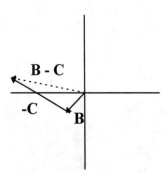

8. <u>Graphically adding and subtracting vectors allows us to visualize a physical situation better than analytically adding and subtracting does.</u>

9.

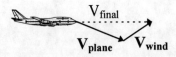 

10. This problem is a straightforward example of using Equations (4.6) and (4.7):

$$A_x = (3.2 \text{ m}) \cdot \cos(15°) = \underline{3.1 \text{ m}}$$

$$A_y = (3.2 \text{ m}) \cdot \sin(15°) = \underline{0.83 \text{ m}}$$

11. In this problem, we are given the x- and y-components of a vector and are asked to calculate its magnitude and direction. To get the magnitude, we use Equation (4.2):

$$\text{Magnitude} = \sqrt{V_x^2 + V_y^2} = \sqrt{(1.7 \text{ m/sec})^2 + (-1.1 \text{ m/sec})^2} = 2.0 \text{ m/sec}$$

To get the angle, we start with Equation (4.3):

$$\theta = \tan^{-1}\left(\frac{V_y}{V_x}\right) = \tan^{-1}\left(\frac{-1.1 \frac{\text{m}}{\text{sec}}}{1.7 \frac{\text{m}}{\text{sec}}}\right) = \tan^{-1}(-0.647) = -33°$$

We aren't necessarily finished yet, however. We have to determine which region of the Cartesian coordinate system that the vector is in. Since the x-component is positive and the y-component is negative, the vector is to the right and below the origin, which means that the vector is in region IV. According to our rules, we add 360 degrees to the result of Equation (4.3) when the vector is in region IV, so 327° is the proper angle. Thus, <u>the vector has magnitude of 2.0 m/sec and direction of 327°</u>.

12. The first step in adding vectors analytically is to break both vectors down into their components:

$$A_x = (31.1 \text{ m}) \cdot \cos(60.0°) = 15.6 \text{ m}$$

$$A_y = (31.1 \text{ m}) \cdot \sin(60.0°) = 26.9 \text{ m}$$

$$B_x = (11.4 \text{ m}) \cdot \cos(290.0°) = 3.90 \text{ m}$$

$$B_y = (11.4 \text{ m}) \cdot \sin(290.0°) = -10.7 \text{ m}$$

Now that we have the individual components, we can add them like numbers.

$$C_x = A_x + B_x = 15.6 \text{ m} + 3.90 \text{ m} = 19.5 \text{ m}$$

$$C_y = A_y + B_y = 26.9 \text{ m} + -10.7 \text{ m} = 16.2 \text{ m}$$

Now that we have the components to our answer, we can use Equations (4.2) and (4.3) to give us the magnitude and direction of the answer. To get the magnitude of this vector, we use Equation (4.2):

$$\text{Magnitude} = \sqrt{C_x^2 + C_y^2} = \sqrt{(19.5 \text{ m})^2 + (16.2 \text{ m})^2} = 25.4 \text{ m}$$

To find the direction of the vector, we use Equation (4.3):

$$\theta = \tan^{-1}\left(\frac{C_y}{C_x}\right) = \tan^{-1}\left(\frac{16.2 \text{ m}}{19.5 \text{ m}}\right) = 39.7°$$

Since the x and y-components are both positive, the vector is in the region I of the Cartesian coordinate plane, so we need not adjust it at all. Thus, the sum of vectors **A** and **B** has a magnitude of 25.4 m at a direction of 39.7°.

13. In this two-dimensional problem, we are given two displacement vectors followed by the boy scout, and we are asked to come up with his final displacement vector. The final vector must be the sum of the two. In the end, then, we simply have to add these two vectors. The first step is to break both vectors down into their components:

$$A_x = (1.8 \text{ miles}) \cdot \cos(150.0°) = -1.6 \text{ miles}$$

$$A_y = (1.8 \text{ miles}) \cdot \sin(150.0°) = 0.90 \text{ miles}$$

$$B_x = (3.2 \text{ miles}) \cdot \cos(350.0°) = 3.2 \text{ miles}$$

$$B_y = (3.2 \text{ miles}) \cdot \sin(350.0°) = -0.56 \text{ miles}$$

Now that we have the individual components, we can add them like numbers.

$$C_x = A_x + B_x = -1.6 \text{ miles} + 3.2 \text{ miles} = 1.6 \text{ miles}$$

$$C_y = A_y + B_y = 0.90 \text{ miles} + -0.56 \text{ miles} = 0.34 \text{ miles}$$

Now that we have the components to our answer, we can use Equations (4.2) and (4.3) to give us the magnitude and direction of the answer. To get the magnitude of this vector, we use Equation (4.2):

$$\text{Magnitude} = \sqrt{C_x^2 + C_y^2} = \sqrt{(1.6 \text{ miles})^2 + (0.34 \text{ miles})^2} = 1.6 \text{ miles}$$

To find the direction of the vector, we use Equation (4.3):

$$\theta = \tan^{-1}\left(\frac{C_y}{C_x}\right) = \tan^{-1}\left(\frac{0.34 \text{ miles}}{1.6 \text{ miles}}\right) = 12°$$

Since the both components of vector **C** are positive, the vector is in region I of the Cartesian coordinate plane, indicating that we need not adjust it at all. So the scout's final displacement is <u>1.6 miles at 12°</u>.

14. Since the boat will be carried by the river's current, the actual velocity of the boat will be the vector sum of the boat and the current. To do this, we first split them up into their components:

$$V_{boat_x} = (15 \text{ mph}) \cdot \cos(130.0°) = -9.6 \text{ mph}$$

$$V_{boat_y} = (15 \text{ mph}) \cdot \sin(130.0°) = 11 \text{ mph}$$

$$V_{current_x} = (2.1 \text{ mph}) \cdot \cos(200.0°) = -2.0 \text{ mph}$$

$$V_{current_y} = (2.1 \text{ mph}) \cdot \sin(200.0°) = -0.72 \text{ mph}$$

Now we add the components together to get the components of the final velocity:

$$V_{final_x} = -9.6 \text{ mph} + -2.0 \text{ mph} = -11.6 \text{ mph}$$

$$V_{final_y} = 11 \text{ mph} + -0.72 \text{ mph} = 1.0 \times 10^1 \text{ mph}$$

With these components, we can get the magnitude and direction of the final velocity:

$$\text{Magnitude} = \sqrt{V_{final_x}^2 + V_{final_y}^2} = \sqrt{(-11.6 \text{ mph})^2 + (1.0 \times 10^1 \text{ mph})^2} = 15 \text{ mph}$$

To find the direction of the vector, we use Equation (4.3):

$$\theta = \tan^{-1}\left(\frac{V_{final_y}}{V_{final_x}}\right) = \tan^{-1}\left(\frac{1.0 \times 10^1 \text{ mph}}{-11.6 \text{ mph}}\right) = -41°$$

Since the x-component is negative and the y-component is positive, we are in region II and we need to add 180.0° to it. The boat's final velocity, then, is $\underline{15 \text{ mph at } 139°}$

15. The final velocity of the plane will be the vector sum of the plane's velocity and the wind's velocity.

$$V_{plane_x} = (200.0 \text{ mph}) \cdot \cos(90.0°) = 0$$

$$V_{plane_y} = (200.0 \text{ mph}) \cdot \sin(90.0°) = 2.00 \times 10^2 \text{ mph}$$

$$V_{wind_x} = (15.0 \text{ mph}) \cdot \cos(315°) = 10.6 \text{ mph}$$

$$V_{wind_y} = (15.0 \text{ mph}) \cdot \sin(315°) = -10.6 \text{ mph}$$

Now we add the components together to get the components of the final velocity:

$$V_{final_x} = 0 \text{ mph} + 10.6 \text{ mph} = 10.6 \text{ mph}$$

$$V_{final_y} = 2.00 \times 10^2 \text{ mph} + -10.6 \text{ mph} = 189 \text{ mph}$$

With these components, we can get the magnitude and direction of the final velocity:

$$\text{Magnitude} = \sqrt{V_{final_x}^2 + V_{final_y}^2} = \sqrt{(10.6 \text{ mph})^2 + (189 \text{ mph})^2} = 189 \text{ mph}$$

To find the direction of the vector, we use Equation (4.3):

$$\theta = \tan^{-1}\left(\frac{V_{final_y}}{V_{final_x}}\right) = \tan^{-1}\left(\frac{189 \text{ mph}}{10.6 \text{ mph}}\right) = 86.8°$$

Since both components are positive, the final vector is in region I of the Cartesian coordinate plane. Therefore, we need not alter the result of Equation (4.3). The final velocity, then, is $\underline{189 \text{ mph at } 86.8°}$.

ANSWERS TO THE MODULE #5 TEST AND EXPERIMENT

1. As with one-dimensional motion, the speed of a projectile when launched will be equal to the speed when it reaches the launch height. Thus, it was launched with a speed of 124 m/sec.

2. When a projectile lands at the same height from which it was launched, it hits its maximum height halfway through its journey. Thus, it will reach its maximum height in 0.22 sec.

3. No. The parabolic motion of a projectile comes from the fact that there is no acceleration in the x-dimension and a strong downward acceleration in the y-dimension. Anything that changes these conditions destroys the parabolic nature of the motion.

4. The projectile must land at the same height from which it was launched.

5. According to the range equation, the range is proportional to the initial speed squared. If the initial speed is doubled, the range will increase by 2^2. Thus, the range increases by a factor of 4.

6. Air resistance would decrease the range, because air resistance inhibits motion.

7. This is like a one-dimensional projectile problem. When it reaches the launch height, its velocity will be equal and opposite the launch velocity, or -12.1 m/sec.

8. The jet airplane and the rocket have engines that change the horizontal and vertical acceleration. The only situation in which horizontal acceleration stays at 0 and the vertical acceleration stays at -9.8 m/sec^2 is a.

9. The range on the moon would be much larger, because the lower gravitational acceleration will result in a longer time in the air.

10. As with all such problems, we need to look for the one-dimensional situations within this two-dimensional problem. Each leg of the journey is an independent one-dimensional problem, because the explorer travels in a straight line in each case. In the first leg of the journey, the velocity is 2.1 m/sec, the acceleration is 0 (because the velocity is constant), and the time is 1.1 hours. After converting time to seconds, Equation (3.19), then, looks like this:

$$x = v_o \cdot t + \frac{1}{2} \cdot a \cdot t^2$$

$$x = (2.1 \frac{m}{\text{sec}}) \cdot (4.0 \times 10^3 \text{ sec}) + \frac{1}{2} \cdot 0 \cdot (4.0 \times 10^3 \text{ sec})^2$$

$$x = 8.4 \times 10^3 \text{ m}$$

Is this our answer? Of course not! This was only the first leg of the journey. We did learn something, however. We learned that the first leg of the journey resulted in a displacement of 8.4 x 10^3 m at an angle of 120.0°. That, then, is where the second leg of the journey begins. If we can figure out the displacement that results from this second leg, we should have enough information to determine the final displacement of the ship.

Well, we can analyze the second leg of the journey the same way that we analyzed the first leg.

$$\mathbf{x} = \mathbf{v}_o \cdot t + \frac{1}{2} \cdot \mathbf{a} \cdot t^2$$

$$\mathbf{x} = (2.1 \frac{m}{\text{sec}}) \cdot (8.3 \times 10^3 \text{ sec}) + \frac{1}{2} \cdot 0 \cdot (8.3 \times 10^3 \text{ sec})^2$$

$$\mathbf{x} = 1.7 \times 10^4 \text{ m}$$

What do we have now? Well, we know that the first leg of the journey resulted in a displacement of 8.4 x 10^3 m at 120.0 degrees, while the second leg resulted in a displacement of 1.7 x 10^4 m at 200.0 degrees. How can we find the total displacement? We can add them in order to get the final displacement.

To add these two displacements, we must first break the vectors down into their components. To make the notation easier, let's call the displacement vector resulting from the first leg of the journey vector **A**. We will therefore call the second vector **B**, and the final displacement vector will be known as **C**.

$$A_x = (8.4 \times 10^3 \text{ m}) \cdot \cos(120.0) = -4.2 \times 10^3 \text{ m}$$

$$A_y = (8.4 \times 10^3 \text{ m}) \cdot \sin(120.0) = 7.3 \times 10^3 \text{ m}$$

$$B_x = (1.7 \times 10^4 \text{ m}) \cdot \cos(200.0) = -1.6 \times 10^4 \text{ m}$$

$$B_y = (1.7 \times 10^4 \text{ m}) \cdot \sin(200.0) = -5.8 \times 10^3 \text{ m}$$

$$C_x = -4.2 \times 10^3 \text{ m} + -1.6 \times 10^4 \text{ m} = -2.0 \times 10^4 \text{ m}$$

$$C_y = 7.3 \times 10^3 \text{ m} + -5.8 \times 10^3 \text{ m} = 1.5 \times 10^3 \text{ m}$$

Now we can determine the magnitude and direction of the final displacement vector:

$$C = \sqrt{C_x^2 + C_y^2} = \sqrt{(-2.0\times 10^4 \text{ m})^2 + (1.5\times 10^3 \text{ m})^2} = 2.0\times 10^4 \text{ m}$$

$$\theta = \tan^{-1}\left(\frac{1.5\times 10^3 \text{ m}}{-2.0\times 10^4 \text{ m}}\right) = -4.3°$$

Based on our rules regarding Equation (4.3), the angle for the final displacement vector is 180.0° + -4.3 = 175.7°. In the end, the ship's final displacement vector is $\underline{2.0 \times 10^4 \text{ m at } \theta = 175.7°}$.

11. The only thing we need to calculate here is the maximum height of the arrow. This means we are concentrating only on the y-dimension. Thus, we had better get the y-component of the initial velocity:

$$V_{o_y} = (112 \frac{\text{ft}}{\text{sec}}) \cdot \sin(40.0°) = 72.0 \frac{\text{ft}}{\text{sec}}$$

We now know the initial velocity in the y-dimension (72.0 ft/sec), the acceleration (-32 ft/sec²), and the final velocity (at the maximum height, the final velocity is 0). From these data, we need to calculate the height. Equation (3.15) will do that:

$$v^2 = v_o^2 + 2ax$$

$$0^2 = (72.0 \frac{\text{ft}}{\text{sec}})^2 + 2 \cdot (-32 \frac{\text{ft}}{\text{sec}^2}) \cdot x$$

$$x = \frac{-(72.0 \frac{\text{ft}}{\text{sec}})^2}{2 \cdot (-32 \frac{\text{ft}}{\text{sec}^2})} = 81 \text{ ft}$$

This tells us that the maximum height the projectiles reach is $\underline{81 \text{ ft}}$.

12. This is a simple application of Equation (5.9). We are given the initial speed (175 m/sec) and the angle (30.0°). From those facts, we are asked to calculate the range of the projectile:

$$\text{Range} = \frac{v_o^2 \cdot \sin 2\theta}{g}$$

$$\text{Range} = \frac{(175 \frac{\text{m}}{\text{sec}})^2 \cdot \sin(2 \cdot 30.0°)}{9.8 \frac{\text{m}}{\text{sec}^2}} = 2.7 \times 10^3 \text{ m}$$

So the gun's range is $\underline{2.7 \times 10^3 \text{ m}}$.

13. In this use of Equation (5.9), we are given the angle (45°) and the range (4,123 ft). We need to determine the necessary initial speed. Thus, we just need to rearrange Equation (5.9) to solve for v_o after we have plugged in the numbers that we know:

$$\text{Range} = \frac{v_o^2 \cdot \sin 2\theta}{g}$$

$$4{,}123 \text{ ft} = \frac{v_o^2 \cdot \sin(2 \cdot 45)}{32 \frac{\text{ft}}{\text{sec}^2}}$$

$$v_o^2 = \frac{4{,}123 \text{ ft} \cdot 32 \frac{\text{ft}}{\text{sec}^2}}{1}$$

$$v_o = 3.6 \times 10^2 \frac{\text{ft}}{\text{sec}}$$

So, the initial velocity is $\underline{3.6 \times 10^2 \text{ ft/sec}}$.

14. In this problem, we need to calculate the displacement in the x-dimension. As usual, however, we need a piece of information from the other dimension before we can do this. We need to use the y-dimension to figure out the time. In that dimension, the initial velocity is given by:

$$V_{o_y} = (20.0 \text{ ft/sec}) \cdot \sin(0.000) = 0$$

We also know that the acceleration is -32 m/sec² and that the displacement in this dimension is -45 ft. This is all we need to use Equation (3.19) to calculate time:

$$x = v_o \cdot t + \frac{1}{2} \cdot a \cdot t^2$$

$$-45 \text{ ft} = (0) \cdot t + \frac{1}{2} \cdot (-32 \frac{\text{ft}}{\text{sec}^2}) \cdot t^2$$

$$t^2 = \frac{2 \cdot 45 \text{ ft}}{32 \frac{\text{ft}}{\text{sec}^2}}$$

$$t = 1.7 \text{ sec}$$

Now that we have time, we can go back and use Equation (3.19) in the x-dimension. To do this, however, we need to calculate the initial velocity in that dimension:

$$V_{o_x} = (20.0 \text{ ft / sec}) \cdot \cos(0.000) = 20.0 \text{ ft / sec}$$

Now we can use Equation (3.19):

$$x = v_o \cdot t + \frac{1}{2} \cdot a \cdot t^2$$

$$x = (20.0 \frac{\text{ft}}{\text{sec}}) \cdot (1.7 \text{ sec}) + \frac{1}{2} \cdot (0) \cdot (1.7 \text{ sec})^2$$

$$x = 34 \text{ ft}$$

So the diver lands <u>34 ft</u> from the cliff.

15. In this problem, of course, we cannot use Equation (5.9), because the arrow lands below the height from which it was launched. We must analyze each dimension, then, and figure out what to use where. We are asked to calculate the initial velocity of the arrow. If we look at the initial velocity in each of the two dimensions, we will quickly see which dimension allows us to calculate that quantity:

$$V_{o_x} = V_o \cdot \cos(0.000) = V_o$$

$$V_{o_y} = V_o \cdot \sin(0.000) = 0$$

So, because the arrow is launched horizontally, the y-component of the velocity is zero. As a result, we can not calculate the initial velocity using the y-dimension. The x-dimension has a non-zero initial velocity, but we do not have enough information yet. In the x-dimension, we know the displacement (118 m) and acceleration (0). If we could find out the time, we could use Equation (3.19) to determine the initial velocity. Unfortunately, we don't have the time yet. Let's see if we can get that from the y-dimension.

In the y-dimension, we know the initial velocity (0), the acceleration (-9.8 m/sec^2), and the displacement (-20.0 m). That's all we need to calculate the time from Equation (3.19).

$$x = v_o \cdot t + \frac{1}{2} \cdot a \cdot t^2$$

$$-20.0 \text{ m} = (0) \cdot t + \frac{1}{2} \cdot (-9.8 \frac{m}{\sec^2}) \cdot t^2$$

$$t^2 = \frac{2 \cdot (-20.0 \text{ m})}{(-9.8 \frac{m}{\sec^2})}$$

$$t = 2.0 \sec$$

Now we know the time it took the arrow to reach the target. At this point, we can go back to the x-dimension to finish the problem:

$$x = v_o \cdot t + \frac{1}{2} \cdot a \cdot t^2$$

$$118 \text{ m} = (V_o) \cdot (2.0 \sec) + \frac{1}{2} \cdot (0) \cdot (2.0 \sec)^2$$

$$V_o = \frac{118 \text{ m}}{2.0 \sec}$$

$$V_o = 59 \frac{m}{\sec}$$

This means that the arrow has an initial speed of 59 m/sec.

Experiment Answer: The times should be identical within the 10% experimental error. You see, in each case we were interested in only one dimension: the y-dimension. When you dropped the rubber band, the initial velocity in the y-dimension was zero and the acceleration was due solely to gravity. When you flipped the rubber band, the initial velocity in the y-dimension was also zero, because the rubber band was shot at an angle of zero degrees. This makes the y-component of the initial velocity zero because the sin of zero is zero. The acceleration in this case was also due solely to gravity. So…in both cases everything in the y-dimension was identical. Thus, the times should be identical as well. In other words, even though the rubber band was moving in two dimensions when you flipped it, there was no difference between the two cases in the dimension of interest. As a result, the times should be identical. This means that the answer to the original question is that the dropped bullet and the fired bullet will hit the ground at the same time.

If your times were not identical within 10%, you most likely determined that it took longer for the rubber band to hit the ground when it was flipped compared to when it was dropped. This is the most common mistake made when performing the experiment. You see, it is difficult for your helper to tell when the rubber band really hits the ground as it is traveling away from him or her. As a result, the timer is often late in stopping the watch. You can partially correct this problem if the timer stands close to where you expect the rubber band to land. That gives him or her a more accurate means of determining when the rubber band does, indeed, hit the ground.

ANSWERS TO THE MODULE #6 TEST

1. <u>First Law - An object at rest or in motion stays in that same state until acted on by an outside force.</u>

<u>Second Law - The sum of the forces applied to an object is equal to the object's mass times its acceleration.</u>

<u>Third Law - For every applied force there is an equal and opposite force.</u>

2. <u>The bicyclist will fall forward off of the bicycle. Because the cyclist had the same velocity as the bicycle, he will continue moving forward once the bicycle stops. The force of friction between him and the bicycle seat will not be great enough to stop that motion. Since the Law of Inertia says an object in motion stays in motion until acted on by an outside force, the bicyclist will stay in motion, falling forward.</u>

3. Since the friction force is directly proportional to the normal force, and since the normal force is equal and opposite of the object's weight, <u>the first will experience twice as much friction as the second.</u>

4. <u>This must be a mass measurement, since only mass has the unit of kg.</u>

5. <u>The force of static friction is greater than that of kinetic friction because the bumps and grooves in the interacting surfaces have time to settle deeply into each other when the object is at rest. Once it's moving, there isn't time for them to nestle as snugly together.</u> (NOTE: IF the student simply says that static friction (or friction coefficient) is greater than kinetic friction (or friction coefficient), give half credit).

6. Force is calculated by taking mass and multiplying by acceleration. Since acceleration has units of distance divided by time squared, the force will have the mass unit (mg) multiplied by the distance unit (cm), divided by the time unit (min) squared. The unit, then, would be $\underline{\frac{cm \cdot mg}{min^2}}$.

7. <u>Newton's Third Law does not inhibit motion, because the equal and opposite forces are not applied on the same object. In order for forces to cancel, they must be applied on the same object.</u>

8. <u>The evidence of the pole-vaulter's force is the fact that the cushion gets crushed. The cushion applies an equal and opposite force on the pole-vaulter, which is evidenced by the fact that the pole-vaulter's downward velocity decreases to zero.</u>

9. This is a simple application of Equation (6.1), as long as we convert g to kg:

$$F = m \cdot a$$

$$F = (0.152 \text{ kg}) \cdot (1.21 \frac{m}{sec^2})$$

$$F = 0.184 \frac{kg \cdot m}{sec^2} = 0.184 \text{ Newtons}$$

The force required is <u>0.184 Newtons</u>.

10. This is a simple application of Equation (6.2) in English units:

$$w = m \cdot g$$

$$w = (34 \text{ slugs}) \cdot (32 \frac{ft}{sec^2})$$

$$w = 1.1 \times 10^3 \frac{slug \cdot ft}{sec^2} = 1.1 \times 10^3 \text{ pounds}$$

The weight is <u>1.1×10^3 pounds</u>.

11. When an object moves from one planet to another, its weight changes but its mass doesn't. We therefore must convert from weight to mass so that we have something which is the same at both places. Since we have the weight on earth, we must use the acceleration due to gravity of earth in order to calculate the mass:

$$w = m \cdot g$$

$$24561 \text{ Newtons} = m \cdot (9.8 \frac{m}{sec^2})$$

$$m = \frac{24561 \frac{kg \cdot m}{sec^2}}{9.8 \frac{m}{sec^2}} = 2.5 \times 10^3 \text{ kg}$$

The mass is the same on the earth and on Jupiter. Thus, we can use this mass and the acceleration due to gravity on Jupiter to determine the weight on Jupiter:

$$w = m \cdot g$$

$$w = (2.5 \times 10^3 \text{ kg}) \cdot (23.2 \frac{m}{\sec^2})$$

$$w = 5.8 \times 10^4 \frac{\text{kg} \cdot m}{\sec^2} = 5.8 \times 10^4 \text{ Newtons}$$

The weight of the ship on Jupiter is $\underline{5.8 \times 10^4 \text{ Newtons}}$.

12. When an object moves from one planet to another, its weight changes but its mass doesn't. We therefore must convert from weight to mass so that we have something which is the same at both places. Since we have the weight on Venus, we must use the acceleration due to gravity of Venus in order to calculate the mass:

$$w = m \cdot g$$

$$1231 \text{ pounds} = m \cdot (28 \frac{ft}{\sec^2})$$

$$m = \frac{1231 \frac{\text{slug} \cdot ft}{\sec^2}}{28 \frac{ft}{\sec^2}} = 44 \text{ slugs}$$

The mass is the same on the earth and on Venus. Thus, we can use this mass and the acceleration due to gravity on earth to determine the weight on earth:

$$w = m \cdot g$$

$$w = (44 \text{ slugs}) \cdot (32 \frac{ft}{\sec^2})$$

$$w = 1.4 \times 10^3 \frac{\text{slug} \cdot ft}{\sec^2} = 1.4 \times 10^3 \text{ pounds}$$

On earth, it weighs $\underline{1.4 \times 10^3 \text{ pounds}}$.

13. This is a simple application of Equation (6.3). Remember, pounds means force, so we already have the weight of the car. The normal force is equal and opposite the weight, so we can just stick it right into the equation. Since we are asked about the friction that keeps the statue from moving, we are calculating static friction and must therefore use μ_s.

$$f = \mu \cdot F_n$$

$$f = (0.32) \cdot (567 \text{ pounds})$$

$$f = 1.8 \times 10^2 \text{ pounds}$$

The frictional force is $\underline{1.8 \times 10^2 \text{ pounds}}$.

14. We must start all problems like this one by looking at all of the forces that come into play:

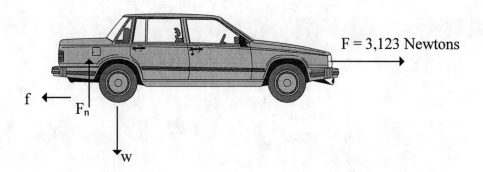

The forces are the weight (w) of the car, the normal force (F_n), the pushing force (3123 Newtons), and friction (f). To calculate the car's acceleration, we need to know all forces acting on the car in the horizontal direction, because that's where the motion takes place. When we know all of those forces, we can sum them up and set them equal to the mass times the acceleration. We already know one of the two forces (3123 Newtons), but we need to calculate friction. To do that, we must first calculate the normal force:

$$w = m \cdot g$$

$$w = (721 \text{kg}) \cdot (9.8 \, \frac{m}{\sec^2})$$

$$w = 7.1 \times 10^3 \, \frac{\text{kg} \cdot \text{m}}{\sec^2} = 7.1 \times 10^3 \text{ Newtons}$$

Since the normal force counteracts the weight, we know that it also has a magnitude of 7.1×10^2 N. We can now calculate friction. To do this, we will use the coefficient of kinetic friction, because the problem tells us to make the calculation for when the car is already moving:

$$f = \mu \cdot F_n$$

$$f = (0.30) \cdot (7.1 \times 10^3 \text{ Newtons})$$

$$f = 2.1 \times 10^3 \text{ Newtons}$$

Now we can *finally* sum up the horizontal forces. We need to take the directions of the force into account explicitly in our summation, so I will say that the 3123 Newtons with which the engine is pushing is a positive force (since it is pointed to the right) and that the frictional force is therefore negative:

$$F - f = m \cdot a$$

$$3123 \text{ Newtons} - 2.1 \times 10^3 \text{ Newtons} = (721 \text{ kg}) \cdot a$$

$$a = \frac{1.0 \times 10^3 \; \frac{\text{kg} \cdot \text{m}}{\text{sec}^2}}{721 \text{ kg}} = 1.4 \; \frac{\text{m}}{\text{sec}^2}$$

The car accelerates at 1.4 m/sec². The significant figures here are a little tricky, so don't count the problem totally wrong if the student doesn't get them right. On the top of the fraction, you have numbers being subtracted. The numbers are 3123 and 2100. When you subtract them, you get 1000. In subtracting, we keep the lowest precision, which is in the 2100 number. That precision is to the hundreds place, so that's why we round it to 1.0 x 10³, because the zero after the decimal is in the hundreds place.

15. As with all such problems, we need to examine all of the forces at play before we try to solve the problem:

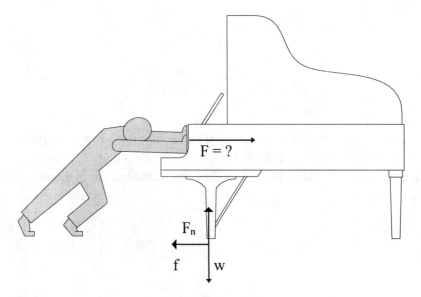

To determine the force needed, we will have to sum up the forces in the horizontal dimension. Before we do that, however, we need to determine friction. We start by determining the normal force. The piano's weight is 1451 pounds. We know that it's weight because its unit is pounds. Since the normal force counteracts weight, this means that the normal force is also 1451 pounds. That's all we need to get started:

$$f = \mu \cdot F_n$$

$$f = (0.24) \cdot (1451 \text{ pounds})$$

$$f = 3.5 \times 10^2 \text{ pounds}$$

Now that we have the frictional force, we can sum up the forces in the horizontal direction and determine the force with which the mover must push. When we sum up the forces, we set them equal to mass times acceleration. In this case, however, the acceleration is zero (constant velocity). Thus, the sum of the forces must equal zero:

$$F - f = 0$$

$$F - 3.5 \times 10^2 \text{ pounds} = 0$$

$$F = 3.5 \times 10^2 \text{ pounds}$$

To keep the wagon moving at a constant velocity, then, the boy must pull with a force of $\underline{3.5 \times 10^2 \text{ pounds}}$.

ANSWERS TO THE MODULE #7 TEST

1. Yes. The object is in dynamic equilibrium.

2. Yes. The object is in dynamic rotational equilibrium.

3. In order to be in static rotational equilibrium, the sum of the torques acting on an object must be zero. In addition, it can have no rotational motion.

4. Torque is calculated by taking force and multiplying by lever arm. Because the second plumber is twice as strong as the first, the force he uses will be twice as large. Because his wrench is twice as long, the lever arm will also be twice as large. The torque the second plumber exerts, then, will be 4 times as strong.

5. Situation A, because the force is applied perpendicular to the lever arm.

6. In situation B, there are no horizontal forces. In situation A, however, one string has a horizontal tension component to the right and the other must fight that with a horizontal tension component to the left. This adds tension, making tension the greatest in situation A.

7. Torque can always be expressed as any force unit times any distance unit. Thus, pound·inch is also an acceptable torque unit.

8. If we look at the forces acting on the baby, there are three of them. Gravity is pulling down on the baby, while the strings pull him up from both sides. Thus, the force diagram looks like this:

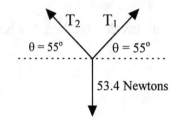

Since the baby is not falling (hopefully), he is in static equilibrium, so the sum of the forces in each dimension must be zero. So, let's split the force vectors up into their components. For T_1, the angle is 55 degrees, but it is not 55 degrees for T_2. In order to properly define the angle for vectors, you must start from the positive x-axis and move counterclockwise. Thus, the properly defined angle for T_2 is 125 degrees. The properly defined angle for the weight vector is 270 degrees. The vectors therefore split up in the following way:

$$T_{1x} = T_1 \cdot \cos(55) = 0.57 \cdot T_1$$

$$T_{1y} = T_1 \cdot \sin(55) = 0.82 \cdot T_1$$

$$T_{2x} = T_2 \cdot \cos(125) = -0.574 \cdot T_2$$

$$T_{2y} = T_2 \cdot \sin(125) = 0.819 \cdot T_2$$

$$w_x = (53.4 \text{ Newtons}) \cdot \cos(270.0) = 0$$

$$w_y = (53.4 \text{ Newtons}) \cdot \sin(270.0) = -53.4 \text{ Newtons}$$

Now that we have everything split up into x and y-components, we can sum up the forces in each dimension and make sure that they equal zero. In the x-dimension:

$$(0.57) \cdot T_1 + (-0.574) \cdot T_2 + 0 = 0$$

Although we can't solve for anything here, we can at least rearrange the equation:

$$T_1 = T_2$$

Now we can go to the y-dimension:

$$(0.82) \cdot T_1 + (0.819) \cdot T_2 + -53.4 \text{ Newtons} = 0$$

This equation also has two unknowns in it, but we can use the fact that T_1 and T_2 are equal (which we learned in the y-dimension) in order to replace T_2 with T_1:

$$(0.82) \cdot T_1 + (0.819) \cdot T_1 + -53.4 \text{ Newtons} = 0$$

$$T_1 = \frac{53.4 \text{ Newtons}}{1.64} = 32.6 \text{ Newtons}$$

Since T_1 and T_2 are equal, this tells us that <u>the tension in each of the strings is 32.6 Newtons</u>.

9. The force diagram in this problem is pretty simple:

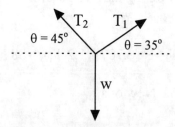

Now we need to split these vectors up into their components:

$$T_{1x} = T_1 \cdot \cos(35) = (0.82) \cdot T_1$$

$$T_{1y} = T_1 \cdot \sin(35) = (0.57) \cdot T_1$$

$$T_{2x} = T_2 \cdot \cos(135) = (-0.707) \cdot T_2$$

$$T_{2y} = T_2 \cdot \sin(135) = (0.707) \cdot T_2$$

$$w_x = (w) \cdot \cos(270.0) = 0$$

$$w_y = (w) \cdot \sin(270.0) = -w$$

Now let's sum up these forces in each dimension and set the sum equal to zero. When we do that, I will just substitute in the weight of the plane for "w." Now remember, the problem gives us mass (9.0 kg), which we must convert to weight (88 Newtons).

$$(0.82) \cdot T_1 + (-0.707) \cdot T_2 = 0$$

Although we cannot solve this equation because it has two variables, we probably realize that we will get another equation in the y dimension, and we will have to solve the equations simultaneously. Therefore, we might as well go ahead and solve for one variable in this equation in terms of the other, to get us started down that route. It doesn't matter which variable we solve for, so I choose to solve for T_2 in terms of T_1:

$$T_2 = \frac{(0.82) \cdot T_1}{0.707} = 1.2 \cdot T_1$$

Now we can go to the y dimension, sum up the forces, and set the sum equal to zero:

$$(0.57) \cdot T_1 + (0.707) \cdot T_2 + - w = 0$$

We can substitute the expression for T_2 that we got from the x-dimension, and we can also put in the weight we calculated earlier:

$$(0.57) \cdot T_1 + (0.707) \cdot (1.2 \cdot T_1) - 88 \text{ Newtons} = 0$$

$$T_1 = 62 \text{ Newtons}$$

We can now use that number to go back to our equation we got from the x-dimension and calculate T_2:

$$T_2 = (1.2) \cdot T_1 = (1.2) \cdot (62 \text{ Newtons}) = 74 \text{ Newtons}.$$

The two strings, then, have a tension of 62 Newtons and 74 Newtons.

10. Equation (7.1) tells us:

$$\tau = F_\perp \cdot r = (456 \text{ Newtons}) \cdot (0.15 \text{ m}) = \underline{68 \text{ Newton} \cdot \text{meters}}$$

11. Each hand generates a torque of

$$\tau = F_\perp \cdot r = (120.0 \text{ Newtons}) \cdot (0.31 \text{ m}) = \underline{37 \text{ Newton} \cdot \text{meters}}$$

Each of the torques moves the wheel in the same direction (clockwise), thus, they add just as if they were one-dimensional vectors pointed in the same direction. The total torque, then, is 74 Newton·meters.

12. To turn a screw, you need torque. Thus, whichever plumber generates more torque is the one that will be most likely to turn the screw.

Plumber #1: $\tau = F_\perp \cdot r = (345 \text{ Newtons}) \cdot (0.45 \text{ m}) = \underline{1.6 \times 10^2 \text{ Newton} \cdot \text{meters}}$

Plumber #2: $\tau = F_\perp \cdot r = (545 \text{ Newtons}) \cdot (0.25 \text{ m}) = \underline{1.4 \times 10^2 \text{ Newton} \cdot \text{meters}}$

The first can exert more torque and is therefore more likely to succeed.

13. To balance the see-saw, all torques must sum up to zero. Since we always ignore the weight of the see-saw in these problems, there are only two torques. The ones caused by the 35 kg (3.4 x 10^2 Newton) girl and the 41 kg (4.0 x 10^2 Newton) boy. The girl's torque is negative because it makes the see-saw turn counterclockwise. The boy's torque is positive because it turns the see-saw clockwise.

$$\tau_{girl} = -(3.4 \times 10^2 \text{ Newtons}) \cdot (1.54 \text{ meters}) = -5.2 \times 10^2 \text{ Newton} \cdot \text{meters}$$

$$\tau_{boy} = (4.0 \times 10^2 \text{ Newtons}) \cdot (x)$$

Adding these torques up and setting them equal to zero:

$$-5.2 \times 10^2 \text{ Newton} \cdot \text{meters} + (4.0 \times 10^2 \text{ Newtons}) \cdot x = 0$$

$$x = 1.3 \text{ meters}$$

To balance the bar, the boy must be 1.3 meters from the fulcrum

14. The force diagram looks like this:

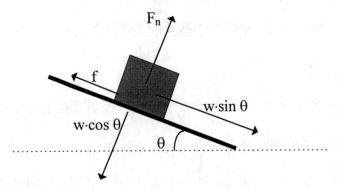

Now we have two forces to consider. One of them is friction, so we might as well go to the perpendicular dimension and determine what the friction is. In that dimension, the sum of the forces needs to equal zero, because there is no movement in that dimension:

$$F_n + -w \cdot \cos\theta = 0$$

$$F_n = (4.4 \times 10^2 \text{ Newtons}) \cdot \cos(41) = 3.3 \times 10^2 \text{ Newtons}$$

Since we now have the normal force, we can calculate friction:

$$f = \mu_k \cdot F_n$$

$$f = (0.45) \cdot (3.3 \times 10^2 \text{ Newtons}) = 1.5 \times 10^2 \text{ Newtons}$$

Now that we have the frictional force, we can finally look at the dimension that contains the motion:

$$f + -w \cdot \sin\theta = m \cdot \mathbf{a}$$

$$1.5 \times 10^2 \text{ Newtons} + -(4.4 \times 10^2 \text{ Newtons}) \cdot \sin(41) = (45 \text{ kg}) \cdot \mathbf{a}$$

$$\mathbf{a} = -3.1 \text{ m/sec}^2$$

The acceleration down the ramp is <u>3.1 m/sec²</u>.

15. When dealing with multiple objects, you have to consider them individually. The force diagram for the rear box looks like this:

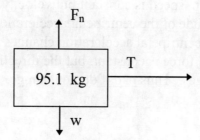

because the tension in the string is the only horizontal force acting on the box at the rear. Thus, this one force equals the box's mass times its acceleration:

$$T = (95.1 \text{ kg}) \cdot a$$

There are two unknowns in this equation, so we need to look at the other box for a second equation. This box has two horizontal forces acting on it: the tension in the string and the child's pull:

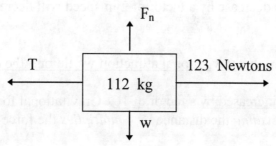

These forces must also sum to the mass of this box times the acceleration:

$$-T + 123 \text{ Newtons} = (112 \text{ kg}) \cdot a$$

This equation also has two unknowns, but we can substitute the expression for T that we found using the other box:

$$-(95.1 \text{ kg}) \cdot a + 123 \text{ Newtons} = (112 \text{ kg}) \cdot a$$

$$a = 0.594 \text{ m/sec}^2$$

Now this is not the answer. The question asks for the tension in the string. So we have to take this acceleration and put it back into the equation we got from the first box. That will give us the tension:

$$T = (95.1 \text{ kg}) \cdot a = (95.1 \text{ kg}) \cdot (0.594 \text{ m/sec}^2) = 56.5 \text{ Newtons}$$

So the string has a tension of 56.5 Newtons.

ANSWERS TO THE MODULE #8 TEST

1. In uniform circular motion, speed is constant but velocity changes because the direction of travel changes. The magnitude of the centripetal acceleration stays constant, but its direction continually changes, so the centripetal acceleration changes as well. In the same way, the magnitude of the centripetal force is constant, but the direction of the force vector, and hence the centripetal force itself changes. Thus <u>b, d, f</u> do not change.

2. 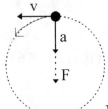 NOTE: The absolute lengths of these vectors are not important.

3. <u>Yes</u>. Centripetal force is necessary for *all* circular motion.

4. <u>Acceleration decreases by a factor of 4</u>. Remember, the centripetal acceleration depends on the speed squared. Thus, a decrease by a factor of 2 in speed will decrease centripetal acceleration by 2^2, or 4.

5. Since there is no friction, <u>the gravitational attraction would pull the objects together</u>.

6. <u>The gravitational force increased by a factor of 16</u>. Gravitational force depends on the inverse of the distance squared. *Dividing* the distance by 4 *multiplies* the force by 4^2.

7. <u>The gravitational force exerted by the sun on the planets is centripetal, allowing for circular motion</u>.

8. In order to hold a satellite in a circular orbit, the gravitational attraction between the earth and the satellite must act as a centripetal force. Based on the equation for centripetal force, the larger the velocity, the greater the centripetal force. Thus, the first object needs the greatest centripetal force. Since the satellites have the same mass, the only way to increase the centripetal force on the first satellite is to move it closer to the earth. As a result, <u>the second</u> satellite orbits at a higher altitude.

9. This problem is a straightforward application of Equation (8.2):

$$a_c = \frac{v^2}{r}$$

$$a_c = \frac{(25.0 \ \frac{m}{sec})^2}{31.0 \ m} = 20.2 \ \frac{m}{sec^2}$$

The car, then, needs a centripetal acceleration of $\underline{20.2 \ m/sec^2}$.

10. Since we know the radius of the circle and the car's speed, we can determine how long it takes for the car to travel around the circle once. That's the period:

$$\text{circumference} = 2 \cdot \pi \cdot (31.0 \text{ m}) = 195 \text{ m}$$

That's the distance traveled in one orbit. Now we can calculate the period:

$$\text{distance} = \text{rate} \cdot \text{time}$$

$$\text{time} = \frac{195 \text{ m}}{25.0 \frac{\text{m}}{\text{sec}}} = 7.80 \text{ sec}$$

Now that we know the period, calculating frequency is a snap:

$$f = \frac{1}{T} = \frac{1}{7.80 \text{ sec}} = 0.128 \frac{1}{\text{sec}}$$

The period is 7.80 sec, and the frequency is 0.128 Hz.

11. In order to solve problem, we must first develop an expression for the friction that exists between the tires and the road. The force of friction depends on the normal force (which for flat surfaces is just the weight of the car) times the coefficient of friction. We do not know the coefficient of friction or the mass of the car, so we cannot get a number for the frictional force. We can, however, develop an expression:

$$f = \mu_k \cdot F_n$$

$$f = (\mu_k) \cdot (m) \cdot (9.8 \frac{\text{meters}}{\text{sec}^2})$$

Now, even though we don't have a number for the frictional force, let's go ahead and put the expression we have for it into Equation (8.1). The expression we have for the frictional force represents the maximum possible centripetal force available. If we put that into Equation (8.1), we can use it to determine the maximum speed that the car can have:

$$F_c = \frac{m \cdot v^2}{r}$$

$$(\mu_k) \cdot (m) \cdot (9.8 \frac{\text{m}}{\text{sec}^2}) = \frac{m \cdot (13 \frac{\text{m}}{\text{sec}})^2}{23.1 \text{ meters}}$$

Since mass appears on both sides of the equation, it cancels out. Thus, it doesn't matter that we don't know the mass of the car, because it cancels out of the final equation. Now we can simply solve for μ_k:

$$(\mu_k) \cdot (\cancel{m}) \cdot (9.8 \frac{\text{meters}}{\text{sec}^2}) = \frac{\cancel{m} \cdot (13 \frac{\text{meters}}{\text{sec}})^2}{23.1 \text{ meters}}$$

$$\mu_k = \frac{(13 \frac{\cancel{\text{meters}}}{\text{sec}})^2}{(9.8 \frac{\cancel{\text{meter}}}{\text{sec}^2}) \cdot 23.1 \cancel{\text{ meter}}}$$

$$\mu_k = 0.75$$

So the coefficient of kinetic friction must be at least $\underline{0.75}$, or the car will slip off the curve!

12. This uses Equation (8.3):

$$F_g = \frac{G \cdot m_1 \cdot m_2}{r^2}$$

Since the problem said that the masses were equal, however, we can substitute "m" for both "m_1" and "m_2":

$$F_g = \frac{G \cdot m \cdot m}{r^2}$$

$$12.1 \text{ Newtons} = \frac{(6.67 \times 10^{-11} \frac{\text{Newton} \cdot \text{meters}^2}{\text{kg}^2}) \cdot m^2}{(0.15 \text{ meters})^2}$$

$$m^2 = \frac{(0.15 \cancel{\text{meters}})^2 \cdot (12.1 \cancel{\text{Newtons}})}{(6.67 \times 10^{-11} \frac{\cancel{\text{Newton}} \cdot \cancel{\text{meters}^2}}{\text{kg}^2})}$$

$$m = 6.4 \times 10^4 \text{ kg}$$

Each object has a mass of $\underline{6.4 \times 10^4 \text{ kg}}$.

13. This is a standard satellite problem:

$$F_g = F_c$$

$$\frac{G \cdot m_{sun} \cdot m_{Venus}}{r^2} = \frac{m_{Venus} \cdot v^2}{r}$$

$$\frac{(6.67 \times 10^{-11} \frac{\text{Newton} \cdot \text{m}^2}{\text{kg}^2}) \cdot (2.0 \times 10^{30} \text{ kg})}{1.1 \times 10^{11} \text{ m}} = v^2$$

$$v = 3.5 \times 10^4 \ \frac{\text{m}}{\text{sec}}$$

Venus orbits the sun with a speed of $\underline{3.5 \times 10^4 \text{ m/sec}}$.

14. To figure out how long it takes one of those rocks to make one orbit around the Saturn (its orbital period, in other words), we need to determine how fast it travels. To do this, we must set the gravitational force equal to the centripetal force. Since the rock is traveling in a circle, the mass term in Equation (8.1) refers to the mass of the rock.

$$F_g = F_c$$

$$\frac{G \cdot m_{Saturn} \cdot m_{rock}}{r^2} = \frac{m_{rock} \cdot v^2}{r}$$

$$\frac{(6.67 \times 10^{-11} \frac{\text{Newton} \cdot \text{m}^2}{\text{kg}^2}) \cdot (5.7 \times 10^{26} \text{ kg})}{6.1 \times 10^7 \text{ m}} = v^2$$

$$v = 2.5 \times 10^4 \ \frac{\text{m}}{\text{sec}}$$

This, of course, is not our answer. To calculate period, we need to determine how far the rock travels in one orbit. Given the orbital radius, this is no problem:

$$\text{circumference} = 2 \cdot \pi \cdot (6.1 \times 10^7 \text{ m}) = 3.8 \times 10^8 \text{ m}$$

That's the distance traveled in one orbit. Now we can calculate the period:

$$\text{distance} = \text{rate} \cdot \text{time}$$

$$\text{time} = \frac{3.8 \times 10^8 \text{ m}}{2.5 \times 10^4 \frac{\text{m}}{\text{sec}}} = 1.5 \times 10^4 \text{ sec}$$

This tells us that it takes the rock $\underline{1.5 \times 10^4 \text{ sec (4.2 hours)}}$ to travel once around Saturn.

15. There is only one orbital radius that will allow any given satellite speed. To determine the radius, we simply set the gravitational force equal to the centripetal force, as we always have:

$$F_g = F_c$$

$$\frac{G \cdot m_{earth} \cdot m_{satellite}}{r^2} = \frac{m_{satellite} \cdot v^2}{r}$$

$$\frac{(6.67 \times 10^{-11} \frac{\text{Newton} \cdot \text{m}^2}{\text{kg}^2}) \cdot (5.98 \times 10^{24} \text{ kg})}{(9.4 \times 10^2 \frac{\text{m}}{\text{sec}})^2} = r$$

$$r = 4.5 \times 10^8 \text{ m}$$

Now that we know the radius, we can calculate the distance that the satellite travels in one orbit:

$$\text{circumference} = 2 \cdot \pi \cdot (4.5 \times 10^8 \text{ m}) = 2.8 \times 10^9 \text{ m}$$

That's the distance traveled in one orbit. Now we can calculate the period:

$$\text{distance} = \text{rate} \cdot \text{time}$$

$$\text{time} = \frac{2.8 \times 10^9 \text{ m}}{9.4 \times 10^2 \frac{\text{m}}{\text{sec}}} = 3.0 \times 10^6 \text{ sec}$$

This tells us that it takes the satellite $\underline{3.0 \times 10^6 \text{ sec (35 days)}}$ to travel once around the earth.

ANSWERS TO THE MODULE #9 TEST

1. <u>Energy is simply the ability to do work. It needn't actually be done. Work, however, involves action. Something must move for work to be done. They have in common the same unit, the Joule.</u>

2. <u>Potential Energy - Energy that is stored, ready to do work</u>

   <u>Kinetic Energy - Energy in motion</u>

3. <u>The potential chemical energy in the fuel is transformed into kinetic heat energy. That heat energy is further transformed into the kinetic mechanical energy in the motion of the water vapor. This is then transformed into the kinetic mechanical energy of the turning fans. Finally, that kinetic mechanical energy is transformed into kinetic electrical energy.</u>

4. <u>He could pull directly horizontally.</u> This would make all of the force parallel to the direction of motion.

5. <u>No</u>, because friction always opposes motion.

6. <u>This roller coaster will never make it up the second hill.</u> When it starts out, its total energy is determined by its potential energy, which is determined by the height of the hill. In order to climb the next hill, it would need more energy than the first hill provided.

7. <u>Each of the workers did the same amount of work.</u> Remember, work is simply the force applied times the distance. Since they took the same number of bricks and placed them on the same table, they exerted the same force over the same distance. <u>The first exerted more power</u>, however, because power is work divided by time. Since he got done in half the time, his power is calculated by dividing by a smaller number. Thus, his power will be larger.

8. <u>The energy is transformed into heat</u>.

9. This is a straightforward application of Equation (9.1):

$$W = F \cdot x$$

$$175 \text{ J} = (F) \cdot (0.75 \text{ m})$$

$$F = \frac{175 \frac{\text{kg} \cdot \text{m}^2}{\text{sec}^2}}{0.75 \text{ m}} = 2.3 \times 10^2 \frac{\text{kg} \cdot \text{m}}{\text{sec}^2}$$

Assuming he pushed directly horizontally, he exerted a force of <u>2.3 x 10² Newtons</u>.

10. When the car sits at the top of the first hill, it has no kinetic energy, because it is not moving. Thus, all of its energy is potential. Since we know the height of the hill, we can calculate that potential energy:

$$PE = m \cdot g \cdot h$$

$$PE = (0.025 \text{ kg}) \cdot (9.8 \frac{m}{\sec^2}) \cdot (2.3 \text{ m}) = 0.56 \text{ J}$$

Now remember, this is also the *total* energy that the car has, because it has no kinetic energy at the top of the hill. As a result, when the car hits the top of the next hill, the total energy is still 0.56 J. Since the car is still moving, at this point, it also has kinetic energy. All we know is that the sum of the potential energy and kinetic energy must be 0.56 J. Since we have the height of the hill, however, we can calculate the potential energy:

$$PE = m \cdot g \cdot h$$

$$PE = (0.025 \text{ kg}) \cdot (9.8 \frac{m}{\sec^2}) \cdot (0.91 \text{ m}) = 0.22 \text{ J}$$

Now that we have the potential energy, we can use Equation (9.4) to calculate the kinetic energy:

$$TE = PE + KE$$

$$0.56 \text{ J} = 0.22 \text{ J} + KE$$

$$KE = 0.56 \text{ J} - 0.22 \text{ J} = 0.34 \text{ J}$$

Now that we have the kinetic energy, we can finally calculate the speed:

$$KE = \frac{1}{2} \cdot m \cdot v^2$$

$$0.34 \text{ J} = \frac{1}{2} \cdot (0.025 \text{ kg}) \cdot v^2$$

$$v = \sqrt{\frac{2 \cdot 0.34 \text{ J}}{0.025 \text{ kg}}} = 5.2 \frac{m}{\sec}$$

At the top of the second hill, the car is traveling at 5.2 m/sec.

11. This problem is much like the problem above, but the cyclist actually has some kinetic energy at the top of the hill. Thus, we need to add potential energy and kinetic energy in order to get the total energy. Since we know the height of the hill, we can calculate the potential energy:

$$PE = m \cdot g \cdot h$$

$$PE = (89 \text{ kg}) \cdot (9.8 \frac{m}{\sec^2}) \cdot (14.1 \text{ m}) = 1.2 \times 10^4 \text{ J}$$

Since we have the speed, we can also calculate the kinetic energy:

$$KE = \frac{1}{2} \cdot m \cdot v^2$$

$$KE = \frac{1}{2} \cdot (89 \text{ kg}) \cdot (9.7 \frac{m}{\sec})^2$$

$$KE = 4.2 \times 10^3 \text{ J}$$

The total energy, then, is the sum of the two:

$$TE = KE + PE = 1.2 \times 10^4 \text{ J} + 4.2 \times 10^3 \text{ J} = 1.6 \times 10^4 \text{ J}$$

Since the cyclist coasts, the total energy does not change. As a result, when the cyclist hits the top of the next hill, the total energy is still 1.6×10^4 J. We know, then, that the sum of the potential energy and kinetic energy must be 1.6×10^4 J. Since we have the height of the hill, we can calculate the potential energy:

$$PE = m \cdot g \cdot h$$

$$PE = (89 \text{ kg}) \cdot (9.8 \frac{m}{\sec^2}) \cdot (2.3 \text{ m}) = 2.0 \times 10^3 \text{ J}$$

Now that we have the potential energy, we can use Equation (9.4) to calculate the kinetic energy:

$$TE = PE + KE$$

$$1.6 \times 10^4 \text{ J} = 2.0 \times 10^3 \text{ J} + KE$$

$$KE = 1.6 \times 10^4 \text{ J} - 2.0 \times 10^3 \text{ J} = 1.4 \times 10^4 \text{ J}$$

Now that we have the kinetic energy, we can finally calculate the speed:

$$KE = \frac{1}{2} \cdot m \cdot v^2$$

$$1.4 \times 10^4 \text{ J} = \frac{1}{2} \cdot (89 \text{ kg}) \cdot v^2$$

$$v = \sqrt{\frac{2 \cdot 1.4 \times 10^4 \text{ J}}{89 \text{ kg}}} = 18 \, \frac{\text{m}}{\text{sec}}$$

At the top of the second hill, the cyclist is traveling at <u>18 m/sec</u>.

12. When the carpenter gives the block a kick, he gives it kinetic energy:

$$KE = \frac{1}{2} \cdot m \cdot v^2$$

$$KE = \frac{1}{2} \cdot (2.3 \text{kg}) \cdot (5.1 \, \frac{\text{m}}{\text{sec}})^2$$

$$KE = 3.0 \times 10^1 \text{ J}$$

Since the floor is level, we can ignore potential energy. Thus, the total energy of the book is 30 J. In order to come to a halt, the book must lose all of that energy to friction. This means that friction must do 30 J of work. Since we know the coefficient of kinetic friction, we can calculate the force with which friction works:

$$f = \mu \cdot F_n$$

$$f = (0.67) \cdot (2.3 \text{ kg}) \cdot (9.8 \, \frac{\text{m}}{\text{sec}^2}) = 15 \text{ Newtons}$$

Now that we know the work done and the force, we can calculate the distance:

$$W = F \cdot x$$

$$3.0 \times 10^1 \text{ J} = (15 \text{ Newtons}) \cdot x$$

$$x = \frac{3.0 \times 10^1 \, \cancel{\text{Newton}} \cdot m}{15 \, \cancel{\text{Newtons}}} = 2.0 \, m$$

The block, therefore, slides 2.0 m before coming to a halt.

13. Since the ball is sitting still at the top of the hill, its total energy is the same as its potential energy. Since we have the height and the mass, we can calculate it:

$$PE = m \cdot g \cdot h$$

$$PE = (0.345 \, kg) \cdot (9.8 \frac{m}{sec^2}) \cdot (3.1 \, m) = 1.0 \times 10^1 \, J$$

If friction were not involved, how much total energy would the ball have at the bottom of the hill? It would have 10 J, all of which would be kinetic. Since friction is working on the ball, however, its total energy will be less. How much less? Well, we have the speed and mass of the ball, so we can calculate it:

$$KE = \frac{1}{2} \cdot m \cdot v^2$$

$$KE = \frac{1}{2} \cdot (0.345 \, kg) \cdot (4.2 \frac{m}{sec})^2$$

$$KE = 3.0 \, J$$

Now, since the ball has no height, this energy is also its total energy. Thus, during the fall, the box went from a total energy of 1.0 x 10¹ J to a total energy of 3.0 J. There were 7 J lost. Where did those 7 J go? Friction converted them into heat. The only way friction could do that was by working, so friction did 7 J of work.

14. We are given the Wattage of the bulb and the time it burns. Using Equation (9.5), we can therefore calculate the energy that was used:

$$P = \frac{W}{t}$$

$$101 \, Watts = \frac{W}{1.80 \times 10^3 \, sec}$$

$$W = (101 \frac{J}{sec}) \cdot (1.80 \times 10^3 \text{ sec}) = 1.82 \times 10^5 \text{ J}$$

The light consumed $\underline{1.82 \times 10^5 \text{ J}}$ of energy.

15. Power is work divided by time. Since we have force and distance, we can calculate work:

$$W = F \cdot x$$

$$W = (121 \text{ Newtons}) \cdot (15.1 \text{ m}) = 1.8 \times 10^3 \text{ J}$$

Once we convert time into seconds, we can plug this work into Equation (9.5):

$$P = \frac{W}{t}$$

$$P = \frac{1.8 \times 10^3 \text{ J}}{1.4 \times 10^2 \text{ sec}} = 13 \text{ Watts}$$

The machine must have at least $\underline{13 \text{ Watts}}$ of power.

ANSWERS TO THE MODULE #10 TEST

1. Momentum and velocity must always point in the same direction. Thus, c contains the only possible velocity and momentum vectors.

2. The first man has more mass.

3. The speeds were the same, but the directions of the velocity vectors were different. Since direction is part of momentum, their momentum vectors were not equal.

4. When you hit an airbag, the airbag flexes with you. This significantly increases the time it takes to stop you. With that large time interval, you can be stopped with a small force. The small force reduces injuries.

5. When air escapes, molecules speed out the back of the balloon. Thus, they have momentum. In order to conserve momentum, the balloon must speed off in the opposite direction.

6. Angular momentum is calculated by multiplying mass, radius, and velocity. Thus, any unit for angular momentum must have a mass unit times a distance unit times a velocity unit. The only units with that in this problem are: $\frac{g \cdot mm^2}{min}$ and $\frac{in^2 \cdot slug}{hr}$

7. No. Cats use their tails to generate angular momentum. To conserve that momentum, the body swings the other way. With no tail, there is nothing to get the angular momentum going.

8. This is a simple application of Equation (10.1)

$$p = m \cdot v$$

$$p_{final} = (0.125 \text{kg}) \cdot (5.4 \frac{m}{sec}) = 0.68 \frac{kg \cdot m}{sec}$$

Since momentum and velocity always have the same direction, the momentum is $0.68 \frac{kg \cdot m}{sec}$ at 300 degrees southeast. **NOTE:** This problem is wrong without the direction.

9. In order to stop the car, the tree must exert an impulse. We can determine it by figuring out the change in momentum. We know that the final momentum is zero, but we need to calculate the initial momentum.

$$p = m \cdot v$$

$$p = (365 \text{ kg}) \cdot (21.1 \frac{m}{sec}) = 7.7 \times 10^3 \frac{kg \cdot m}{sec}$$

There is also a direction here. The problem says that the car is traveling horizontally, so we will assume that positive motion is in the direction of the car's motion.

Now we know the change in momentum. If the car was traveling with a momentum of $7.7 \times 10^3 \frac{kg \cdot m}{sec}$ and its momentum changed to zero, the change in momentum (final minus initial) is $-7.7 \times 10^3 \frac{kg \cdot m}{sec}$. The negative sign tells us the direction of the momentum change. Now we can calculate the force.

$$F = \frac{\Delta p}{\Delta t}$$

$$F = \frac{-7.7 \times 10^3 \frac{kg \cdot m}{sec}}{0.22 \ sec} = -3.5 \times 10^4 \frac{kg \cdot m}{sec^2}$$

The tree, then, exerts a force of <u>-3.5 x 10⁴ Newtons or 3.5 x 10⁴ Newtons opposite the initial velocity of the car.</u>

10. In order to turn the ball around and send it flying opposite of the direction in which the pitcher threw it, the bat must exert an impulse. We can determine it by figuring out the change in momentum. We know mass and the velocities, so we can calculate initial and final momenta.

Initial:
$$p = m \cdot v$$

$$p = (0.2500 \ kg) \cdot (38 \frac{m}{sec}) = 9.5 \frac{kg \cdot m}{sec}$$

Final:
$$p = m \cdot v$$

$$p = (0.2500 \ kg) \cdot (-52 \frac{m}{sec}) = -13 \frac{kg \cdot m}{sec}$$

The negative sign simply tells us that the ball is moving opposite of the direction in which it was originally thrown. Now we can calculate the change in momentum:

$$\Delta p = p_{final} - p_{initial} = -13 \frac{kg \cdot m}{sec^2} - 9.5 \frac{kg \cdot m}{sec^2} = -23 \frac{kg \cdot m}{sec^2}$$

Now we know the change in momentum as well as the force applied, so we can calculate the time interval.

$$F = \frac{\Delta p}{\Delta t}$$

$$-201 \text{Newtons} = \frac{-23 \frac{\text{kg} \cdot \text{m}}{\text{sec}}}{\Delta t}$$

$$\Delta t = \frac{-23 \frac{\text{kg} \cdot \text{m}}{\text{sec}}}{-201 \frac{\text{kg} \cdot \text{m}}{\text{sec}^2}} = 0.11 \text{ sec}$$

The ball and bat are in contact for 0.11 sec.

11. In order to determine the velocity, we must calculate the total momentum of the system both before and after the gun was fired. Since the Law of Momentum Conservation says that they must equal each other, we can build the following equation:

$$(m_{gun} \cdot v_{gun} + m_{bullet} \cdot v_{bullet})_{before} = (m_{gun} \cdot v_{gun} + m_{bullet} \cdot v_{bullet})_{after}$$

Since both the gun and the shell are stationary, both momenta are equal to zero before the shot. We have all of the rest of the information in the equation except for the velocity of the bullet after the shot was fired. We can therefore solve for it.

$$0 + 0 = (2.34 \text{ kg}) \cdot (5.2 \frac{\text{m}}{\text{sec}}) + (0.095 \text{ kg}) \cdot (v_{bullet})$$

$$0 = 12 \frac{\text{kg} \cdot \text{m}}{\text{sec}} + (0.095 \text{ kg}) \cdot v_{bullet}$$

$$v_{bullet} = \frac{-12 \frac{\text{kg} \cdot \text{m}}{\text{sec}}}{0.095 \text{ kg}} = -1.3 \times 10^2 \frac{\text{m}}{\text{sec}}$$

The negative sign simply indicates that the bullet travels in the opposite direction as compared to the gun. Thus, the gun shoots its bullets at 1.3×10^2 m/sec opposite of the recoil velocity.

12. Since the sum of the forces acting on our system (the skater and the ball) is zero, we can simply calculate the momenta before and after the throw and set them equal to each other. Since

the skater catches the ball after the collision, the ball and skater afterwards can be thought of as one object, moving with a single velocity and the sum of their masses.

$$(m_{ball} \cdot v_{ball} + m_{skater} \cdot v_{skater})_{before} = (m_{ball} + m_{skater}) \cdot v_{after}$$

The skater stands still initially, so her initial velocity (and therefore initial momentum) is zero.

$$(8.0 \text{ kg}) \cdot (3.7 \frac{m}{\sec}) + 0 = (60.0 \text{ kg} + 8.0 \text{ kg}) \cdot v_{skater}$$

$$v_{skater} = \frac{(8.0 \text{ kg}) \cdot (3.7 \frac{kg \cdot m}{\sec})}{68.0 \text{ kg}} = 0.44 \frac{m}{\sec}$$

The lack of a negative sign simply means that the skater travels in the same direction as the ball at 0.44 m/sec.

13. In this problem, the object under consideration (the train car) changes its mass. Since mass is a part of momentum, this affects the car's momentum. Assuming momentum is conserved, we can develop the following equation:

$$m_{empty} \cdot v_{empty} = m_{full} \cdot v_{full}$$

We are told the mass and velocity of the car when it is empty. In addition, we are also told the velocity of the car when it is full. Thus, we can use our equation to calculate the total mass of the car and coal once the train car is filled.

$$(975 \text{ kg}) \cdot (3.1 \frac{m}{\sec}) = (m_{full}) \cdot (1.2 \frac{m}{\sec})$$

$$m_{full} = \frac{(975 \text{ kg}) \cdot (3.1 \frac{m}{\sec})}{1.2 \frac{m}{\sec}} = 2.5 \times 10^3 \text{ kg}$$

This is not the answer, however. This is the *total* mass of the train car plus the coal. To get the mass of the coal, we subtract them;

$$\text{Mass of coal} = 2.5 \times 10^3 \text{ kg} - 975 \text{ kg} = 1.5 \times 10^3 \text{ kg}$$

The train car was filled with 1.5 x 10³ kg of coal.

14. Angular momentum is given by Equation (10.11):

$$L = m \cdot v \cdot r$$

$$L = (7.36 \times 10^{22} \text{ kg}) \cdot (1.00 \times 10^3 \frac{\text{m}}{\text{sec}}) \cdot (3.8 \times 10^8 \text{ m}) = 2.8 \times 10^{34} \frac{\text{kg} \cdot \text{m}^2}{\text{sec}}$$

The angular momentum is $\underline{2.8 \times 10^{34} \frac{\text{kg} \cdot \text{m}^2}{\text{sec}}}$.

15. Once the toy starts twirling, it has angular momentum. When the string is adjusted, the radius of the motion changes, but the angular momentum cannot, due to the Law of Angular Momentum Conservation. As a result, the angular momentum before the radius changes must be the same as the angular momentum after it changes:

$$(m \cdot v \cdot r)_{before} = (m \cdot v \cdot r)_{after}$$

Since the mass of the toy does not change, it is the same on both sides of the equation, so it cancels out. We know everything else but the speed afterwards, so we can solve for it:

$$\cancel{m} \cdot (3.9 \frac{\text{meters}}{\text{sec}}) \cdot (0.45 \text{ meters}) = \cancel{m} \cdot (v_{after}) \cdot (0.98 \text{ meters})$$

$$v_{after} = \frac{(3.9 \frac{\cancel{\text{meters}}}{\text{sec}}) \cdot (0.45 \text{ meters})}{9.8 \cancel{\text{meters}}} = 1.8 \frac{\text{meters}}{\text{sec}}$$

The ball's new speed is $\underline{1.8 \text{ m/sec}}$.

ANSWERS TO THE MODULE #11 TEST

1. In simple harmonic motion, the restoring force must be directly proportional to the displacement from equilibrium. Thus, the equation in b results in simple harmonic motion.

2. In a mass/spring system, the period is independent of the amplitude. Thus, the new period will also be 1.1 seconds.

3. At the equilibrium position, where potential energy is zero

4. At the amplitude, where the displacement from equilibrium is greatest

5. Displace it a small distance. Pendulums cease to exhibit simple harmonic motion at large displacements from equilibrium.

6. The periods are the same because the period of a pendulum depends only on the acceleration due to gravity and the length of the pendulum.

7. Because Equation (11.20) tells us that the period of a pendulum is proportional to the square root of the length, the first has a shorter period than the second.

8. The amplitude is defined as the maximum displacement from equilibrium. This is the same as the initial displacement from equilibrium. Thus, the initial displacement is 12.1 cm.

9. This is a simple application of Equation (11.11):

$$T = 2 \cdot \pi \sqrt{\frac{m}{k}}$$

$$T = 2 \cdot \pi \sqrt{\frac{34.5 \text{ kg}}{12.1 \frac{\text{Newtons}}{\text{m}}}} = 2 \cdot \pi \sqrt{\frac{34.5 \text{ kg}}{12.1 \frac{\text{kg} \cdot \text{m}}{\text{sec}^2}}}$$

$$T = 10.6 \text{ sec}$$

The period is 10.6 seconds.

10. This is problem uses Equation (11.1).

$$\mathbf{F} = -\mathbf{k} \cdot \mathbf{x}$$

$$F = -2.9 \frac{\text{Newtons}}{\text{m}} \cdot (-0.151 \text{m})$$

$$\mathbf{F = 0.44 \text{ Newtons}}$$

Note that the force is positive. Since we assumed that the displacement was negative, the fact that the force is positive just means that the force which the mass experiences is opposite the direction of the displacement.

Why does the spring exert this restoring force? Because the weight of the mass pulls down on the spring. Since the restoring force stops the mass from moving, it must be equal and opposite of the mass's weight. Thus, the weight of the mass is 0.44 Newtons. We can use that to calculate the mass.

$$W = m \cdot g$$

$$0.44 \text{ Newtons} = m \cdot (9.8 \frac{\text{m}}{\text{sec}^2})$$

$$m = \frac{0.44 \frac{\text{kg} \cdot \text{m}}{\text{sec}^2}}{9.8 \frac{\text{m}}{\text{sec}^2}} = 0.045 \text{ kg}$$

The mass of the object is <u>0.045 kg</u>.

11. When the fish hangs on the spring, it stretches downwards by 9.2 cm. Thus, the displacement (**x**) is -0.092 m, after converting to standard units. Using the mass, we can calculate the weight of the fish. This must equal the restoring force, so that the fish does not keep falling. Thus, with this information, we can calculate the spring constant:

$$W = m \cdot g = (5.61 \text{ kg}) \cdot (9.8 \frac{\text{m}}{\text{sec}^2}) = 55 \text{ Newtons}$$

$$\mathbf{F = -k \cdot x}$$

$$55 \text{ Newtons} = -k \cdot (-0.092 \text{ m})$$

$$k = \frac{55 \text{ Newtons}}{0.092 \text{ m}} = 6.0 \times 10^2 \frac{\text{Newtons}}{\text{m}}$$

The spring constant, therefore, is 6.0 x 10² Newtons/meter. With this information, we can now calculate the period.

$$T = 2 \cdot \pi \sqrt{\frac{m}{k}}$$

$$T = 2 \cdot \pi \sqrt{\frac{5.61 \text{ kg}}{6.0 \times 10^2 \frac{\text{Newtons}}{\text{m}}}} = 2 \cdot \pi \sqrt{\frac{5.61 \text{ kg}}{6.0 \times 10^2 \frac{\text{kg} \cdot \text{m}}{\text{sec}^2 \cdot \text{m}}}}$$

$$T = 0.61 \text{ sec}$$

The period is 0.61 seconds.

12. When the mass is displaced to 30.0 cm, it has no speed, so its kinetic energy is zero. Its potential energy is calculated from Equation (11.12):

$$PE = \frac{1}{2} \cdot k \cdot x^2$$

$$PE = \frac{1}{2} \cdot (5.11 \frac{\text{Newtons}}{\text{m}}) \cdot (0.300 \text{ m})^2$$

$$PE = 0.230 \text{ Newton} \cdot \text{m} = 0.230 \text{ Joules}$$

Since there is no kinetic energy, 0.230 J is also the total energy.

13. The maximum speed will occur when all of the 0.230 J is converted to kinetic energy:

$$KE = \frac{1}{2} \cdot m \cdot v^2$$

$$0.230 \text{ J} = \frac{1}{2} \cdot (45.0 \text{ kg}) \cdot v^2$$

$$v = \sqrt{\frac{2 \cdot \left(0.230 \frac{\text{kg} \cdot \text{m}^2}{\text{sec}^2}\right)}{45.0 \text{ kg}}} = 0.101 \frac{\text{m}}{\text{sec}}$$

The maximum speed is 0.101 m/sec.

14. When it is 10.0 cm from equilibrium, it will have both potential and kinetic energy. Given the displacement, we can calculate the potential energy:

$$PE = \frac{1}{2} \cdot k \cdot x^2$$

$$PE = \frac{1}{2} \cdot (5.11 \, \frac{\text{Newtons}}{\text{m}}) \cdot (0.100 \, \text{m})^2$$

$$PE = 0.0256 \, \text{Newton} \cdot \text{m} = 0.0256 \, \text{Joules}$$

To get the kinetic energy, we can subtract the potential energy from the total:

$$KE = TE - PE = 0.230 \, J - 0.0256 \, J = 0.204 \, J$$

This allows us to calculate the speed:

$$KE = \frac{1}{2} \cdot m \cdot v^2$$

$$0.204 \, J = \frac{1}{2} \cdot (45.0 \, \text{kg}) \cdot v^2$$

$$v = \sqrt{\frac{2 \cdot \left(0.204 \, \frac{\text{kg} \cdot \text{m}^2}{\text{sec}^2}\right)}{45.0 \, \text{kg}}} = 0.0952 \, \frac{\text{m}}{\text{sec}}$$

The speed at a displacement of 10.0 cm is <u>0.0952 m/sec</u>.

15. The pendulum's period depends on length and gravitation acceleration. Since we know the length, we can calculate the gravitational acceleration:

$$T = 2 \cdot \pi \cdot \sqrt{\frac{\ell}{g}}$$

$$2.8 \, \text{sec} = 2 \cdot \pi \cdot \sqrt{\frac{0.75 \, \text{m}}{g}}$$

$$(2.8 \text{ sec})^2 = 4 \cdot \pi^2 \cdot \left(\frac{0.75 \text{ m}}{g} \right)$$

$$g = \frac{4 \cdot \pi^2 \cdot (0.75 \text{ m})}{(2.8 \text{ sec})^2} = 3.8 \ \frac{\text{m}}{\text{sec}^2}$$

This gravitation acceleration indicates that the planet is Mars.

ANSWERS TO THE MODULE #12 TEST

1. The oscillation and propagation are perpendicular to one another, so the wave is <u>transverse</u>.

2. You see the image in a flat mirror because your brain extrapolates the light rays. Thus, the image is <u>virtual</u>.

3. The speed of light increases with decreasing density. Thus, light travels faster in <u>substance B</u>.

4. Sound speed increases with increasing density. Thus, sound travels faster in <u>substance A</u>.

5. Pitch is determined by frequency. A higher pitch means a higher frequency, which means a smaller wavelength. Thus, <u>the tenor</u> has the longest wavelength because he has the lower pitch.

6. Index of refraction increases with increasing density. Thus, <u>substance B</u> is more dense.

7. <u>The angle of incidence equals the angle of reflection</u>

8. <u>The first singer's sound waves have a larger amplitude</u>. Wavelengths and frequencies affect pitch, while amplitude affects volume.

9. We can relate frequency and wavelength from Equation (12.1). Remember, the speed of light is 3.0×10^8 m/sec.

$$f = \frac{v}{\lambda}$$

$$1.2 \times 10^3 \, \frac{1}{\text{sec}} = \frac{3.0 \times 10^8 \, \frac{\text{m}}{\text{sec}}}{\lambda}$$

$$\lambda = \frac{3.0 \times 10^8 \, \frac{\text{m}}{\text{sec}}}{1.2 \times 10^3 \, \frac{1}{\text{sec}}} = 2.5 \times 10^5 \, \text{m}$$

The wavelength is $\underline{2.5 \times 10^5 \, \text{m}}$.

10. Wavelength and frequency are related by Equation (12.1). To use this, however, we need to determine the speed of the waves. For sound, we use Equation (12.2):

$$v = (331.5 + 0.60 \cdot T) \, \frac{\text{m}}{\text{sec}}$$

$$v = (331.5 + 0.60 \cdot 25) \, \frac{\text{m}}{\text{sec}} = 347 \, \frac{\text{m}}{\text{sec}}$$

Now we can use Equation (12.2).

$$f = \frac{v}{\lambda}$$

$$f = \frac{347 \, \frac{m}{sec}}{0.512 \, m} = 678 \, \frac{1}{sec}$$

The frequency is <u>678 Hz</u>.

11. To determine temperature from sound, we need to determine its speed. We do that with Equation (12.1):

$$f = \frac{v}{\lambda}$$

$$612 \, \frac{1}{sec} = \frac{v}{0.580 \, m}$$

$$v = (612 \, \frac{1}{sec}) \cdot (0.580 \, m) = 355 \, \frac{m}{sec}$$

Now we can use Equation (12.2) to determine temperature:

$$v = (331.5 + 0.60 \cdot T) \, \frac{m}{sec}$$

$$355 \, \frac{m}{sec} = (331.5 + 0.60 \cdot T) \, \frac{m}{sec}$$

$$T = \frac{355 - 331.5}{0.60} = 4.0 \times 10^1$$

The temperature is <u>4.0×10^1 °C</u>.

12. We assume that the light reaches the observer instantaneously; thus, the time delay is the time that it takes for the sound to travel the distance between the observer and where the lightning was formed. To determine that distance, we need to determine the speed:

$$v = (331.5 + 0.60 \cdot T) \, \frac{m}{sec}$$

$$v = (331.5 + 0.60 \cdot 11) \frac{m}{sec} = 338.1 \frac{m}{sec}$$

Now that we know the speed, and we also know that there is no acceleration, we can simply use Equation (3.19) to determine the distance:

$$x = v_o t + \frac{1}{2} a t^2$$

$$x = (338.1 \frac{m}{\text{sec}}) \cdot (2.9 \text{ sec}) + \frac{1}{2} \cdot (0) \cdot (2.9 \text{ sec})^2 = 9.8 \times 10^2 \text{ m}$$

The lightning was formed 9.8×10^2 m away from the observer.

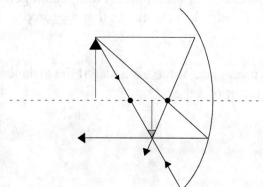

13.   <u>Real, inverted, and reduced</u>

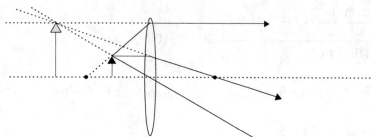

14.   <u>Virtual, upright, and magnified</u>

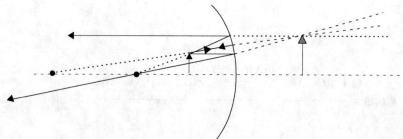

15.   <u>Virtual, upright, and magnified</u>

ANSWERS TO THE MODULE #13 TEST

1. When the charge granted is opposite the charge used, the object was charged by induction.

2. Protons are positive. An excess of protons means a positive charge.

3. The charges would cancel out if the bar was a conductor. Thus, it must be an insulator.

4. We say instantaneous because whatever we calculate is only true for an instant. After that, the particles move under the influence of the force, changing the distance between them, which changes the force.

5. 5.0 N. Remember, the electrostatic force is mutual. Both charges contribute to the force, and they each exert the same amount of force on the other.

6. The electrostatic force is directly proportional to the charge of each object multiplied together. If the charge on one object is halved, then that product is halved. This means that the force decreases by one half.

7. The electrostatic force is inversely proportional to the square of the distance. If the distance decreases by a factor of two, the force increases by a factor of 4.

8. This is an application of Equation (13.1):

$$F = \frac{k \cdot q_1 \cdot q_2}{r^2}$$

$$F = \frac{(9.0 \times 10^9 \, \frac{\text{Newton} \cdot \text{m}^2}{\text{C}^2}) \cdot (2.1 \text{C}) \cdot (9.3 \text{C})}{(0.45 \text{ m})^2} = 8.7 \times 10^{11} \text{ Newtons}$$

The force is 8.7×10^{11} Newtons, and it is attractive because the charges are opposite.

9. This is another application of Equation (13.1):

$$F = \frac{k \cdot q_1 \cdot q_2}{r^2}$$

$$4.3 \times 10^5 \text{ Newtons} = \frac{(9.0 \times 10^9 \, \frac{\text{Newton} \cdot \text{m}^2}{\text{C}^2}) \cdot (3.1 \times 10^{-3} \text{ C}) \cdot q_2}{(1.2 \text{ m})^2}$$

$$q_2 = \frac{(4.3 \times 10^5 \text{ Newtons}) \cdot (1.2 \text{ m})^2}{(9.0 \times 10^9 \frac{\text{Newton} \cdot \text{m}^2}{\text{C}^2}) \cdot (3.1 \times 10^{-3} \text{ C})} = 2.2 \times 10^{-2} \text{ C}$$

Since the force is repulsive, the charges must have the same sign. Thus, the charge is +2.2 x 10^{-2} C.

10. In this problem, we are only worried about the +6.4 mC charge. As a result, we only consider the forces which act on that particular charge. The -5.2 mC charge exerts an attractive force whose magnitude is:

$$F = \frac{k \cdot q_1 \cdot q_2}{r^2}$$

$$F = \frac{(9.0 \times 10^9 \frac{\text{Newtons} \cdot \text{m}^2}{\text{C}^2}) \cdot (6.4 \times 10^{-3} \text{ C}) \cdot (5.2 \times 10^{-3} \text{ C})}{(1.5 \text{ m})^2} = 1.3 \times 10^5 \text{ Newtons}$$

The other force acting on the +6.4 mC charge is the attractive force exerted by the -8.1 mC charge.

$$F = \frac{k \cdot q_1 \cdot q_2}{r^2}$$

$$F = \frac{(9.0 \times 10^9 \frac{\text{Newtons} \cdot \text{m}^2}{\text{C}^2}) \cdot (6.4 \times 10^{-3} \text{ C}) \cdot (8.1 \times 10^{-3} \text{ C})}{(2.0 \text{ m})^2} = 1.2 \times 10^5 \text{ Newtons}$$

The fact that both forces are attractive will give us the directions of the force vectors, making our force diagram look like:

(-5.2 mC) ←— 1.3 x 10^5 Newtons (+6.4 mC) 1.2 x 10^5 Newtons —→ (-8.1 mC)

Since the force vectors both point in the same dimension, we can treat them as one-dimensional vectors. This means we can take care of direction with positives and negatives and then simply add the magnitudes together. Using the convention that vectors pointing to the left are negative, the total force is:

$$\mathbf{F}_{total} = 1.2 \times 10^5 \text{ Newtons} + -1.3 \times 10^5 \text{ Newtons} = -1 \times 10^4 \text{ Newtons}$$

A negative force means that the final vector points to the left. Thus, the final instantaneous electrostatic force is 1×10^4 Newtons to the left.

11. Since we are only interested in the +7.6 mC charge, we only need concern ourselves with forces which act on that charge. The -2.5 mC charge exerts an attractive force on it:

$$F = \frac{k \cdot q_1 \cdot q_2}{r^2}$$

$$F = \frac{(9.0 \times 10^9 \, \frac{\text{Newtons} \cdot \text{m}^2}{\text{C}^2}) \cdot (7.6 \times 10^{-3} \, \text{C}) \cdot (2.5 \times 10^{-3} \, \text{C})}{(0.65 \, \text{m})^2} = 4.0 \times 10^5 \text{ Newtons}$$

The +1.1 mC charge exerts a repulsive force on it:

$$F = \frac{k \cdot q_1 \cdot q_2}{r^2}$$

$$F = \frac{(9.0 \times 10^9 \, \frac{\text{Newtons} \cdot \text{m}^2}{\text{C}^2}) \cdot (7.6 \times 10^{-3} \, \text{C}) \cdot (1.1 \times 10^{-3} \, \text{C})}{(0.65 \, \text{m})^2} = 1.8 \times 10^5 \text{ Newtons}$$

Our force diagram, then, looks like this:

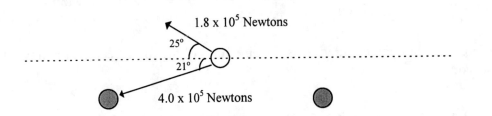

Since these vectors do not point in the same dimension, we will have to add them with trig. First, however, let's define the angles properly. Vector angles are always defined counterclockwise from the +x axis. This means that the first angle is 155°, and the other angle is 201°. Now we can add these vectors:

$$A_x = (1.8 \times 10^5 \text{ Newtons}) \cdot \cos(155°) = -1.6 \times 10^5 \text{ Newtons}$$

$$A_y = (1.8 \times 10^5 \text{ Newtons}) \cdot \sin(155°) = 7.6 \times 10^4 \text{ Newtons}$$

$$B_x = (4.0 \times 10^5 \text{ Newtons}) \cdot \cos(201°) = -3.7 \times 10^5 \text{ Newtons}$$

$$B_y = (4.0 \times 10^5 \text{ Newtons}) \cdot \sin(201°) = -1.4 \times 10^5 \text{ Newtons}$$

$C_x = A_x + B_x = $ -1.6 x 10^5 Newtons + -3.7 x 10^5 Newtons = -5.3 x 10^5 Newtons

$C_y = A_y + B_y = $ 7.6 x 10^4 Newtons + -1.4 x 10^5 Newtons = -6 x 10^4 Newtons

All that's left to do now is convert these x and y components into vector magnitude and direction:

$$\text{Magnitude} = \sqrt{C_x^2 + C_y^2} = \sqrt{(-5.3 \times 10^5 \text{ Newtons})^2 + (-6 \times 10^4 \text{ Newtons})^2} = 5.3 \times 10^5 \text{ Newtons}$$

$$\theta = \tan^{-1}\left(\frac{C_y}{C_x}\right) = \tan^{-1}\left(\frac{-6 \times 10^4 \text{ Newtons}}{-5.3 \times 10^5 \text{ Newtons}}\right) = 6.5°$$

Since both the x and y components of the vector are negative, we know that the vector is in quadrant III. This means that we need to add 180.0° to the angle above to properly define the vector angle. The instantaneous electrostatic force on the 7.6 mC charge, then, is <u>5.3 x 10^5 Newtons at an angle of 186.5°</u>

12. a. Electric field lines come from positive charges and end on <u>negatively charged</u> particles.
    b. There are 4 times as many lines going into B than are leaving A. This means that the charge of B is 4 times the charge of a, or <u>4.0 C</u>.
    c. The least acceleration will occur where the force is the least. In an electric field, the occurs where the line density is the lowest, <u>near particle A, especially to the left</u>.

13.

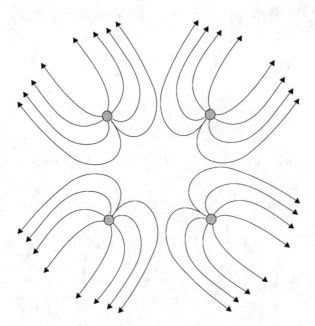

14. The ion in this case has three protons, so its nucleus has a positive charge of 4.8×10^{-19} C (three times the charge of a single proton). Since there is only one electron in the ion, however, we can solve this problem by setting the centripetal force equal to the electrostatic force.

$$\frac{m \cdot v^2}{r} = \frac{k \cdot q_1 \cdot q_2}{r^2}$$

$$v^2 = \frac{k \cdot q_1 \cdot q_2 \cdot \cancel{r}}{r^2 \cdot m} = \frac{k \cdot q_1 \cdot q_2}{r \cdot m} = \frac{(9.0 \times 10^9 \, \frac{\frac{kg \cdot \cancel{m}}{sec^2} \cdot m^2}{\cancel{C^2}}) \cdot (4.8 \times 10^{-19} \, \cancel{C}) \cdot (1.6 \times 10^{-19} \, \cancel{C})}{(5.9 \times 10^{-12} \, \cancel{m}) \cdot (9.1 \times 10^{-31} \, \cancel{kg})}$$

$$v = 1.1 \times 10^7 \, \frac{m}{sec}$$

This electron has a speed of $\underline{1.1 \times 10^7 \text{ m/sec}}$

ANSWERS TO THE MODULE #14 TEST

1. Electrical potential must be in Volts, not Joules.

2. In calculating electrical potential, you use the value of the stationary charge. Thus, the charge of the non-stationary particle is irrelevant. Since the electrical potential is negative, the stationary charge must be negative.

3. Potential energy is the product of electrical potential (whose sign is determined by the sign of the stationary charge) times the charge of the non-stationary particle. If the stationary charge is positive, then the electrical potential is positive. In order to get a positive potential energy, then, the non-stationary particle must also be positive.

4. The stationary positive charge indicates a positive electrical potential. When multiplied by the negative moving charge, the potential energy is negative. As the particles become closer, the electrical potential increases, because electrical potential depends on the inverse of the distance. Thus, as the particles get closer, the potential energy gets more negative. Thus, it decreases.

5. A device that stores charge

6. Regardless of the sign of the particle's charge, a particle gains electrical potential when it moves from the negative to the positive plate. Since this particle travels in the opposite direction, its electrical potential decreases.

7&8.

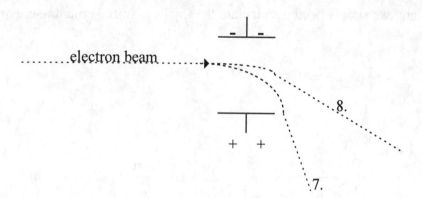

9. The repair person assumes it's the capacitor because the vertical line is caused by the electron beam sweeping up and down the screen. However, since the rest of the screen is black, it makes sense that the electron beam is not being swept horizontally. Thus, the capacitor which controls the horizontal sweep must be the defective one.

10. This is a straightforward application of Equation (14.1), as long as we remember that the electrical potential depends only on the stationary charge.

$$V = \frac{k \cdot q}{r}$$

$$V_{initial} = \frac{\left(9.0 \times 10^9 \frac{N \cdot m^2}{C^2}\right) \cdot (-3.4 \times 10^{-3} \, C)}{0.35 \, m} = 8.7 \times 10^7 \frac{N \cdot m}{C} = 8.7 \times 10^7 \text{ Volts}$$

The electrical potential is $\underline{-8.7 \times 10^7 \text{ Volts}}$.

11. This is a simple application of Equation (14.2)

$$PE = q \cdot V$$

$$-0.45 \, J = (q) \cdot (15 \frac{N \cdot m}{C})$$

$$q = \frac{-0.45 \, J}{15 \frac{N \cdot m}{C}} = -0.030 \, C$$

The units work here because a J is a N·m. The charge, then, $\underline{-0.030 \, C}$.

12. To solve this problem, we simply need to calculate the final and initial potential energy and then subtract the two.

$$V = \frac{k \cdot q}{r}$$

$$V_{initial} = \frac{\left(9.0 \times 10^9 \frac{N \cdot m^2}{C^2}\right) \cdot (6.8 \times 10^{-3} \, C)}{0.75 \, m} = 8.2 \times 10^7 \frac{N \cdot m}{C} = 8.2 \times 10^7 \text{ Volts}$$

$$PE = q \cdot V$$

$$PE_{initial} = (-1.5 \times 10^{-3} \, C) \cdot (8.2 \times 10^7 \frac{N \cdot m}{C}) = -1.2 \times 10^5 \, N \cdot m = -1.2 \times 10^5 \, J$$

$$V = \frac{k \cdot q}{r}$$

$$V_{final} = \frac{\left(9.0 \times 10^9 \frac{N \cdot m^2}{C^2}\right) \cdot (6.8 \times 10^{-3} \, C)}{0.33 \, m} = 1.9 \times 10^8 \frac{N \cdot m}{C} = 1.9 \times 10^8 \text{ Volts}$$

$$PE = q \cdot V$$

$$PE_{final} = (-1.5 \times 10^{-3} \, C) \cdot (1.9 \times 10^8 \frac{N \cdot m}{C}) = -2.9 \times 10^5 \, N \cdot m = -2.9 \times 10^5 \, J$$

Change in PE = $PE_{final} - PE_{initial}$ = $(-2.9 \times 10^5 \, J) - (-1.2 \times 10^5 \, J)$ = $-1.7 \times 10^5 \, J$

The potential decreased by 1.7×10^5 J.

13. We solve this problem just like we solved the problems in Module #9. First, we need to calculate the particle's total energy when it starts. At that point, kinetic energy is zero (because it is at rest), and potential energy can be calculated using Equations (14.1) and (14.2):

$$V = \frac{k \cdot q}{r}$$

$$V_{initial} = \frac{\left(9.0 \times 10^9 \frac{N \cdot m^2}{C^2}\right) \cdot (4.4 \times 10^{-3} \, C)}{1.2 \, m} = 3.3 \times 10^7 \frac{N \cdot m}{C} = 3.3 \times 10^7 \text{ Volts}$$

$$PE = q \cdot V$$

$$PE_{initial} = (1.5 \times 10^{-3} \, C) \cdot (3.3 \times 10^7 \frac{N \cdot m}{C}) = 5.0 \times 10^4 \, N \cdot m = 5.0 \times 10^4 \, J$$

Now that we know both the kinetic and potential energy, we know the total energy initially in the system:

$$TE_{initial} = KE_{initial} + PE_{initial} = 0 \, J + 5.0 \times 10^4 \, J = 5.0 \times 10^4 \, J$$

That total energy can never change. Thus, if we just calculate the potential energy at the end, we can determine the kinetic energy:

$$V = \frac{k \cdot q}{r}$$

$$V_{final} = \frac{\left(9.0 \times 10^9 \frac{N \cdot m^2}{C^2}\right) \cdot (4.4 \times 10^{-3} \, C)}{2.0 \, m} = 2.0 \times 10^7 \frac{N \cdot m}{C} = 2.0 \times 10^7 \text{ Volts}$$

$$PE = q \cdot V$$

$$PE_{final} = (1.5 \times 10^{-3} \, C) \cdot (2.0 \times 10^7 \frac{N \cdot m}{C}) = 3.0 \times 10^4 \, N \cdot m = 3.0 \times 10^4 \, J$$

Since we know that the total energy still must be 5.4 x 10⁴ J, we can use this fact and the final potential energy to calculate the final kinetic energy:

$$TE_{final} = KE_{final} + PE_{final}$$

$$5.0 \times 10^4 \, J = KE_{final} + 3.0 \times 10^4 \, J$$

$$KE_{final} = 2.0 \times 10^4 \, J$$

Now we can calculate the speed with Equation (9.3):

$$KE = \frac{1}{2} \cdot m \cdot v^2$$

$$2.0 \times 10^4 \, J = \frac{1}{2} \cdot (4.3 \, kg) \cdot v^2$$

$$v = 96 \frac{m}{sec}$$

When the particle has traveled to 2.0 m away from the stationary charge, then, its speed is 96 m/sec.

14. To solve this problem, we need to find out where the kinetic energy of the particle equals zero. That's the point at which the particle stops and turns around. As always, we start by calculating the total energy from the initial conditions. First, we start with the potential energy:

$$V = \frac{k \cdot q}{r}$$

$$V_{initial} = \frac{\left(9.0 \times 10^9 \frac{N \cdot m^2}{C^2}\right) \cdot (1.8 \times 10^{-3} \, C)}{1.4 \, m} = 1.2 \times 10^7 \frac{N \cdot m}{C} = 1.2 \times 10^7 \text{ Volts}$$

$$PE = q \cdot V$$

$$PE_{initial} = (2.7 \times 10^{-3} \, C) \cdot (1.2 \times 10^7 \frac{N \cdot m}{C}) = 3.2 \times 10^4 \, N \cdot m = 3.2 \times 10^4 \, J$$

Since the particle is initially moving, it has kinetic energy as well. We need to calculate it in order to get the total energy:

$$KE = \frac{1}{2} \cdot m \cdot v^2$$

$$KE_{initial} = \frac{1}{2} \cdot (1.5 \, kg) \cdot (202 \, \frac{m}{sec})^2$$

$$KE_{initial} = 3.1 \times 10^4 \, J$$

Now that we know both the kinetic and potential energy, we know the total energy initially in the system:

$$TE_{initial} = KE_{initial} + PE_{initial} = 3.1 \times 10^4 \, J + 3.2 \times 10^4 \, J = 6.3 \times 10^4 \, J$$

This number can never change, even when the particle stops. However, when that happens, we know that the particle has zero kinetic energy. Thus, we can calculate the potential energy from these facts:

$$TE_{final} = KE_{final} + PE_{final}$$

$$6.3 \times 10^4 \, J = 0 \, J + PE_{final}$$

$$PE_{final} = 6.3 \times 10^4 \, J$$

In other words, the particle will stop when the potential energy is 6.3 x 10⁴ J. Now we can use Equations (14.1) and (14.2) backwards in order to get the distance at which this occurs. First, since electrical potential is what depends on distance, we need to turn potential energy into electrical potential:

$$PE = q \cdot V$$

$$6.3 \times 10^4 \text{ J} = (2.7 \times 10^{-3} \text{ C}) \cdot V$$

$$V = \frac{6.3 \times 10^4 \text{ J}}{2.7 \times 10^{-3} \text{ C}} = 2.3 \times 10^7 \text{ Volts}$$

The units work out here because a Joule is a C·V. Now that we have electrical potential, we can finally go to Equation (14.1) and calculate the distance between the charges:

$$V = \frac{k \cdot q}{r}$$

$$2.3 \times 10^7 \text{ Volts} = \frac{\left(9.0 \times 10^9 \frac{N \cdot m^2}{C^2}\right) \cdot (1.8 \times 10^{-3} \text{ C})}{r}$$

$$r = \frac{\left(9.0 \times 10^9 \frac{N \cdot m^2}{C^2}\right) \cdot (1.8 \times 10^{-3} \text{ C})}{2.3 \times 10^7 \frac{N \cdot m}{C}} = 0.70 \text{ m}$$

To make the units work out, I replaced Volt with its equivalent unit. So, we finally see that the particle can travel until it is <u>0.70 m away from the charge</u>.

15. As the proton moves from one plate of the capacitor to the other, it experiences a change in electrical potential. We can use Equation (14.3) to determine the amount of change:

$$V = \frac{Q}{C_q}$$

$$V = \frac{1.7 \times 10^{-2} \text{ C}}{2.4 \times 10^{-6} \text{ F}} = \frac{1.7 \times 10^{-2} \text{ C}}{2.4 \times 10^{-6} \frac{C}{V}} = 7.1 \times 10^3 \text{ V}$$

The fact that the proton moves from the positive plate to the negative one means that this is a decrease in potential, so the change in potential is -7.1×10^3 Volts.

This change in potential, then, will lead to a change in potential energy, given by Equation (14.2):

$$PE = q \cdot V = (1.6 \times 10^{-19} \text{ C}) \cdot (-7.1 \times 10^3 \text{ V}) = -1.1 \times 10^{-15} \text{ J}$$

Since the result of the equation is negative, this tells us that the proton's potential energy decreased. Well, if the potential energy of the proton decreased by 1.1×10^{-15} J, then its kinetic energy must have increased by the same amount. Since the kinetic energy started at zero and increased by 1.1×10^{-15} J, we know that the final kinetic energy of the proton is 1.1×10^{-15} J. We can now use Equation (9.3) to determine the final speed:

$$KE = \frac{1}{2} \cdot m \cdot v^2$$

$$1.1 \times 10^{-15} \text{ J} = \frac{1}{2} \cdot (1.7 \times 10^{-27} \text{ kg}) \cdot v^2$$

$$v = 1.1 \times 10^6 \, \frac{\text{m}}{\text{sec}}$$

The proton, therefore, moves at the speed of $\underline{1.1 \times 10^6 \text{ m/sec}}$.

ANSWERS TO THE MODULE #15 TEST

1. Circuit diagrams use conventional current, which is current that flows from positive to negative.

2. Since the replacement of one bulb allowed current to flow to all bulbs, the bulbs are hooked up in series.

3. Resistance is the friction between electrons and the atoms of the conductor. Since all matter is made of atoms, all conductors will have some resistance.

4. An electrical heater's elements have a certain amount of resistance. This resistance slows down the electrons traveling through the heater, converting the kinetic energy of the electrons into heat energy.

5.

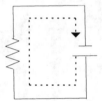

6. There is no break in the conductor in <u>a</u>, so it is the circuit in which the light will work.

7. This is a straightforward application of Equation (15.2):

$$V = I \cdot R = (2.3 \text{ A}) \cdot (56 \text{ }\Omega) = \underline{1.3 \times 10^2 \text{ V}}$$

8. This is a straightforward application of Equation (15.2):

$$V = I \cdot R = (1.7 \text{ A}) \cdot (4.1 \text{ }\Omega) = \underline{7.0 \text{ V}}$$

9. This problem uses Equation (15.4)

$$P = I^2 \cdot R$$

$$98 \text{ Watts} = (2.1 \text{ A})^2 \cdot R$$

$$R = \underline{22 \text{ }\Omega}$$

10. These resistors are hooked up in series because all of the current must go through all of them. Thus, Equation (15.5) gives us the effective resistance:

$$R_{effective} = R_1 + R_2 + R_3 = 271 \, \Omega + 311 \, \Omega + 522 \, \Omega = 366 \, \Omega$$

The effective resistance is <u>1104 Ω</u>.

11. In this circuit, the resistors are hooked up in parallel, because the current has a choice as to which resistor to travel through. As a result, we use Equation (15.6) to determine the effective resistance:

$$\frac{1}{R_{effective}} = \frac{1}{R_1} + \frac{1}{R_2} + \frac{1}{R_3}$$

$$\frac{1}{R_{effective}} = \frac{1}{1.0 \, \Omega} + \frac{1}{2.0 \, \Omega} + \frac{1}{3.0 \, \Omega}$$

$$\frac{1}{R_{effective}} = 1.8 \, \frac{1}{\Omega}$$

$$R_{effective} = 0.56 \, \Omega$$

The effective resistance, then, is <u>0.56 Ω</u>.

12. To determine the power drawn by the circuit, we must determine the current drawn by the circuit.

$$V = I \cdot R$$

$$120.0 \, V = I \cdot (501 \, \Omega)$$

$$I = \frac{120.0 \, V}{501 \, \Omega} = 0.240 \, A$$

This current, then, can be used in Equation (15.3) to calculate the power drawn by the circuit:

$$P = I \cdot V = (0.240 \, A) \cdot (120.0 \, V) = 28.8 \text{ Watts}$$

The circuit draws <u>28.8 Watts</u> of power.

13. To determine the fuse, we need to know the current. For that, we need to get the effective resistance. Since the resistors are in parallel here, we need Equation (15.6) to get the effective resistance:

$$\frac{1}{R_{effective}} = \frac{1}{R_1} + \frac{1}{R_2}$$

$$\frac{1}{R_{effective}} = \frac{1}{11\,\Omega} + \frac{1}{28\,\Omega}$$

$$\frac{1}{R_{effective}} = 0.127\,\frac{1}{\Omega}$$

$$R_{effective} = 7.87\,\Omega$$

We can use this resistance in Equation (15.2) to get the current:

$$V = I \cdot R$$

$$9.0\,V = I \cdot (7.87\,\Omega)$$

$$I = \frac{9.0\,V}{7.87\,\Omega} = 1.1\,A$$

The electrician must use the <u>3 Amp fuse</u>.

14. The circuit must first be simplified:

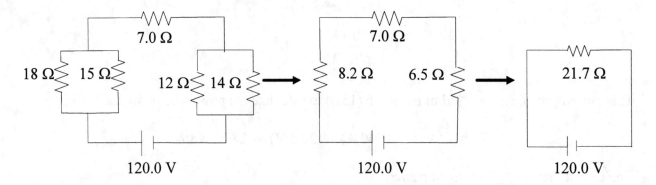

Now we can calculate current, which will then allow us to calculate power.

$$V = I \cdot R$$

$$120.0 \text{ V} = I \cdot (21.7 \text{ }\Omega)$$

$$I = \frac{120.0 \text{ V}}{21.7 \text{ }\Omega} = 5.53 \text{ A}$$

This current, then, can be used in Equation (15.3) to calculate the power drawn by the circuit:

$$P = I \cdot V = (5.53 \text{ A}) \cdot (120.0 \text{ V}) = \underline{6.64 \times 10^2 \text{ Watts}}$$

15. The circuit must first be simplified:

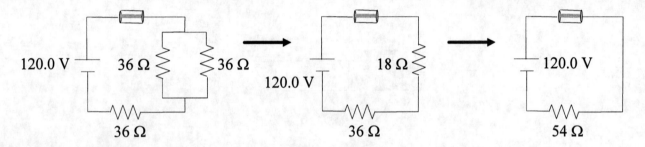

Now we can calculate current:

$$V = I \cdot R$$

$$120.0 \text{ V} = I \cdot (54 \text{ }\Omega)$$

$$I = \frac{120.0 \text{ V}}{54 \text{ }\Omega} = 2.2 \text{ A}$$

The electrician must use the <u>3 Amp fuse</u>.

ANSWERS TO THE MODULE #16 TEST

1. The north pole points in the same direction as the magnetic field lines. The south pole, then, points in the opposite direction:

   →

2.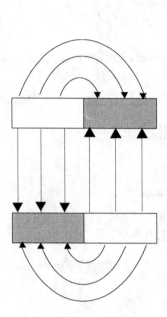

3. There is only one pole. Magnets always have 2 poles.

4. paramagnetic

5. diamagnetic

6. paramagnetic

7. Expose it to a strong magnetic field. This will align its domains.

8

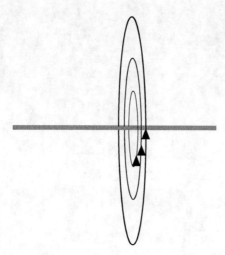

9. When a conductor is exposed to a changing magnetic field, it produces a current. The direction of the magnetic field's change determines the direction of the current.

10. A coil of wires is turned in a magnetic field.

11. steam from boiled water

12. The potential energy of the chemical bonds in the coal is turned into heat energy when the coal is burned. The heat energy then gives the molecules in the water extra kinetic energy so that they can turn into the gas phase. The kinetic energy of the vapor is then turned into the kinetic energy of the turbine, which converts to the kinetic energy of the coil of wires, which finally converts into the kinetic energy of the current traveling through the wires.

13. current which constantly changes direction in a circuit

14. alternating current

15. The rated Voltage is an effective value, since the Voltage is constantly changing.